Managing Projects as Investments

Earned Value to Business Value

Stephen A. Devaux

CRC Press
Taylor & Francis Group
Boca Raton   London   New York

CRC Press is an imprint of the
Taylor & Francis Group, an **Informa** business

First published in paperback 2024

CRC Press
2385 NW Executive Center Drive, Suite 320, Boca Raton FL 33431

and by CRC Press
4 Park Square, Milton Park, Abingdon, Oxon, OX14 4RN

CRC Press is an imprint of Taylor & Francis Group, LLC

© 2015, 2024 by Taylor & Francis Group, LLC

No claim to original U.S. Government works

Version Date: 20140813

Library of Congress Cataloging-in-Publication Data

Devaux, Stephen A., 1949-
    Managing projects as investments : earned value to business value / Stephen A. Devaux.
        pages cm -- (Industrial innovation series)
    Includes bibliographical references and index.
    ISBN 978-1-4822-1270-9 (hardback)
    1. Project management. I. Title.

HD69.P75D478 2014
658.4'04--dc23   2014027319

ISBN: 978-1-4822-1270-9 (hbk)
ISBN: 978-1-03-283674-4 (pbk)
ISBN: 978-0-429-06989-5 (ebk)

DOI: 10.1201/b17496

**Visit the Taylor & Francis Web site at
http://www.taylorandfrancis.com**

**and the CRC Press Web site at
http://www.crcpress.com**

Industrial Innovation Series

Series Editor

Adedeji B. Badiru

Department of Systems and Engineering Management
Air Force Institute of Technology (AFIT) – Dayton, Ohio

PUBLISHED TITLES

Carbon Footprint Analysis: Concepts, Methods, Implementation, and Case Studies,
  Matthew John Franchetti & Defne Apul

Communication for Continuous Improvement Projects, *Tina Agustiady*

Computational Economic Analysis for Engineering and Industry, *Adedeji B. Badiru &
  Olufemi A. Omitaomu*

Conveyors: Applications, Selection, and Integration, *Patrick M. McGuire*

Culture and Trust in Technology-Driven Organizations, *Frances Alston*

Global Engineering: Design, Decision Making, and Communication, *Carlos Acosta, V. Jorge Leon,
  Charles Conrad, & Cesar O. Malave*

Handbook of Emergency Response: A Human Factors and Systems Engineering Approach,
  Adedeji B. Badiru & LeeAnn Racz

Handbook of Industrial Engineering Equations, Formulas, and Calculations, *Adedeji B. Badiru &
  Olufemi A. Omitaomu*

Handbook of Industrial and Systems Engineering, Second Edition *Adedeji B. Badiru*

Handbook of Military Industrial Engineering, *Adedeji B.Badiru & Marlin U. Thomas*

Industrial Control Systems: Mathematical and Statistical Models and Techniques,
  Adedeji B. Badiru, Oye Ibidapo-Obe, & Babatunde J. Ayeni

Industrial Project Management: Concepts, Tools, and Techniques, *Adedeji B. Badiru,
  Abidemi Badiru, & Adetokunboh Badiru*

Inventory Management: Non-Classical Views, *Mohamad Y. Jaber*

Kansei Engineering - 2-volume set
  • Innovations of Kansei Engineering, *Mitsuo Nagamachi & Anitawati Mohd Lokman*
  • Kansei/Affective Engineering, *Mitsuo Nagamachi*

Knowledge Discovery from Sensor Data, *Auroop R. Ganguly, João Gama, Olufemi A. Omitaomu,
  Mohamed Medhat Gaber, & Ranga Raju Vatsavai*

Learning Curves: Theory, Models, and Applications, *Mohamad Y. Jaber*

Managing Projects as Investments: Earned Value to Business Value, *Stephen A. Devaux*

Modern Construction: Lean Project Delivery and Integrated Practices, *Lincoln Harding Forbes &
  Syed M. Ahmed*

Moving from Project Management to Project Leadership: A Practical Guide to Leading Groups,
  R. Camper Bull

Project Management: Systems, Principles, and Applications, *Adedeji B. Badiru*

Project Management for the Oil and Gas Industry: A World System Approach, *Adedeji B. Badiru &
  Samuel O. Osisanya*

Quality Management in Construction Projects, *Abdul Razzak Rumane*

Quality Tools for Managing Construction Projects, *Abdul Razzak Rumane*

Social Responsibility: Failure Mode Effects and Analysis, *Holly Alison Duckworth &
  Rosemond Ann Moore*

PUBLISHED TITLES

Statistical Techniques for Project Control, *Adedeji B. Badiru & Tina Agustiady*

STEP Project Management: Guide for Science, Technology, and Engineering Projects,
    Adedeji B. Badiru

Sustainability: Utilizing Lean Six Sigma Techniques, *Tina Agustiady & Adedeji B. Badiru*

Systems Thinking: Coping with 21st Century Problems, *John Turner Boardman & Brian J. Sauser*

Techonomics: The Theory of Industrial Evolution, *H. Lee Martin*

Triple C Model of Project Management: Communication, Cooperation, Coordination,
    Adedeji B. Badiru

FORTHCOMING TITLES

Cellular Manufacturing: Mitigating Risk and Uncertainty, *John X. Wang*

Essentials of Engineering Leadership and Innovation, *Pamela McCauley-Bush &
    Lesia L. Crumpton-Young*

Technology Transfer and Commercialization of Environmental Remediation Technology,
    Mark N. Goltz

To
my wife Deb

Contents

Preface

Every project needs management for two reasons. The first is to plan and manage the tasks of a project and to coordinate getting the work done. The other is to optimize the value of a project as an investment. Managing projects as investments is the new frontier of project management. This book is focused on introducing the tools and methods for optimizing projects as investments. Texts such as my 1999 *Total Project Control: A Manager's Guide to Integrated Project Planning, Measuring, and Tracking* (Wiley) are aimed primarily at the first function of project management—managing the work—and are written for the "practitioners" of project management: project managers, cost account managers, activity leaders, and all project team members. This book is for both the practitioners and for anyone in a project-driven organization who has responsibility for or is affected by projects.

This includes, but may not be limited to

- Executives in project-driven organizations, who sense that programs and projects are investments and want to shift their focus to understanding them as investments
- Sponsors/senior managers, whose limited budgets must fund projects
- Business area directors, whose portfolios may include multiple projects
- Program managers, whose responsibilities usually include generating value from a coordinated implementation of related project and nonproject efforts
- Customers of projects, especially those projects performed on a contractual basis
- Professionals in contracts departments, responsible for writing or amending requests for quotes (RFQs), requests for proposals (RFPs), and performance contracts for project and subproject vendors and customers
- Business analysis, finance, and accounting professionals, who should be responsible for quantifying and tracking both project cost and project expected and generated value, as well as revenue recognition

- Acquisition professionals, who must understand the significance to project teams of delays in obtaining resources needed on the critical path
- Human resources, who must understand both the urgency of timely recruitment efforts as well as the need for corporate staffing levels, procedures, and incentives that recognize and facilitate critical path progress

The current edition of the Project Management Institute's *Guide to the Project Management Body of Knowledge (PMBOK® Guide)* has a new section, not included in previous editions, on what it terms *business value*, and links it to portfolio, program, and project management. This is recognition, in the most important and widely used project management book, of this new central concept in project management. The major focus of this book is to explain how to use this concept and apply it to plan, measure, track, and optimize the business value of a project. I do that using methods that I have been developing and testing in consulting to project-driven organizations in many industries for more than 25 years.

This book addresses the needs of project managers and team members by providing specific methods that allow them to manifest their value by seeking and finding opportunities to improve project results. But, as important, it shows the executives and all senior managers how to change current processes and metrics that generate inertia and a lack of initiative. Instead, they must encourage processes where projects are measured on the basis of the value they contribute, and where team members have the incentive to seek and recognize value-adding opportunities that increase the strength of the organization.

Because *Total Project Control* was intended for the practitioner, it not only introduced a new approach supported by new techniques, but it delved deeply into each. It explored the different techniques for developing a value breakdown structure (VBS) and for computing critical path drag in network schedules with complex dependencies and lags. Indeed, what was then the brand-new concept of drag and how to compute it is probably the topic that gained that book the most attention and it remains the metric with which I am most identified. In this book, we mention the VBS and discuss critical path drag, but only explore it in simple networks, sufficient to show the importance of the concept and explain its crucial investment corollaries of drag cost and true cost. Anyone with questions about computing drag in complex networks should refer to the previous book, *Total Project Control*.

This book focuses on the why and the how: why we need to start managing projects in the same terms that we use for all other investments and how the unique qualities of a project (work identification, schedule, critical path, resources, cost, and project-specific risk and opportunity) can be analyzed in investment terms.

We venture far beyond the basics into the whole reason why every project investment is made in the first place: because of the benefit/value that the project work is expected to generate. We show the negative effect of leaving project value (and other effects on value such as time) as what in economics would be referred to as *externalities*: items that, because they are not quantified, are measured as zero, even though we know that they can be very important!

We provide techniques for

- Measuring expected value and the value/cost of time. ("Time is money!" wrote Ben Franklin in 1747, a valuable concept for project organizations more than 260 years later.)
- Showing how to plan, track, and analyze all key terms of the investment to provide the best results in the most important metric: value above cost.
- Demonstrating that current popular efforts in project-driven organizations, such as stabilizing inhouse resources at lean staffing levels and emphasizing high utilization rates, are often counterproductive and lead periodically beyond lean and all the way to emaciation in key skills and functions unless such efforts are tempered by a wisdom and flexibility based in investment data.

This book is particularly for those who have both the authority and the responsibility to improve the current condition of project performance in organizations. It shows why ignoring key investment aspects, techniques, and metrics have a deleterious impact on the project team and overall project performance.

Dr. Atul Gawande, creator of the presurgical checklist that has become standard procedure in thousands of hospitals and perhaps saved millions of lives, wrote in his article, "Slow Ideas,"[1] about why certain complex ideas, despite their obvious value, are often slow to take root: "We want frictionless, 'turnkey' solutions to the major difficulties of the world." But as Dr. Gawande points out, not all problems are subject to simple solutions.

2002 Nobel Laureate in Economics Daniel Kahneman[2] notes that "effortful thinking" is painful! People are much more comfortable making snap judgments than considered computations. Yet projects require investment decisions, and these usually demand consideration and calculation of many variables. If organizational processes that mandate the collection and analysis of the necessary investment data are not standardized, "fast thinking" is likely to lead to bad decisions.

At the end of every chapter, a list of specific takeaways provides:

- Simple metrics that allow the funding executive/sponsor or customer to determine, quickly and easily, if the project is on course to provide the anticipated investment value in a timely and cost-efficient manner.
- A procedural map that project management offices (PMOs) can follow and that will lay out for them an agenda of standardized procedures, processes, metrics, and documentation. The implementation of these will both lead the project teams to maximize the investment return and allow the PMO itself to generate and demonstrate the quantitative value it is adding to the organization.
- An array of new techniques for the project manager, the scheduler, and each member of the project team that will permit them to perform their jobs better and to provide greater value to their organizations.

These takeaways are also intended to provide the busy senior manager with a menu of metrics, processes, and techniques that can build appropriate cost–benefit analysis into the organizational culture. Many of these processes can be turned over to the PMO or the individual project manager with instructions to "Make it so." When the project manager believes that she needs more resources, she will have techniques to analyze and justify the expenses in an investment-quantified manner. And when team members are told to "look for opportunities," they will understand where and how to look, and will be able to quantify the cost of the risk or the value of the opportunity in clear monetary terms. This in turn will lead to an appreciation of the value that their analytical and management skills as a team member are adding to the project investment.

Project management has been practiced for a long time, and it is difficult to suggest a fundamental new approach to anything without seeming to criticize those who have previously been doing things in a different way. I want to emphasize that I have the greatest respect for those project management theorists who, decades ago, developed such techniques as the triple constraint project model, work breakdown structures, critical path scheduling and metrics, resource leveling, earned value metrics, and a host of other valuable methods. I am only too aware that modern theorists such as myself "stand on the shoulders of giants," and that my ideas are merely enhancements of techniques and metrics that I inherited from them. Those who have for years and for decades conscientiously applied the traditional toolkit have my greatest respect.

That said, I believe the new focus on investment and business value, when combined with enhancements that are all rooted in the traditional tools (e.g., the value breakdown structure or VBS, critical path drag,

drag and true cost, the cost of leveling with unresolved bottlenecks (or CLUB), and the Devaux's Index of Project Performance (DIPP) progress index or DPI) can bring much greater efficiency and value to those organizations that implement this agenda, and perhaps save lives on those projects where the greatest human value is on the line.

Finally, there are many business leaders who blanch at the sight of mathematical formulas; that is something they would rather delegate to the CFO and the finance department. Yes, there is a mathematical component to project management; as an investment, every project is an economic entity. Critical path scheduling and cost and earned value all require simple calculations. But let me reassure the reader that all the formulas in this book are simple: there is nothing that requires a background or love of arithmetic. And many of my clients and students greatly enjoy the simple computations of critical path method (CPM) scheduling; they say it's like doing Sudokus!

Nevertheless, if the calculations are an irritant, just skip over them. An accountant or a software package can do those calculations. They are included only to illustrate the underlying concepts, which is where the true value of this methodology lies.

Endnotes

1. Atul Gawande, "Slow Ideas." *The New Yorker*, July 29, 2013, p. 4 http://www.newyorker.com/reporting/2013/07/29/130729fa_fact_gawande?currentPage=4
2. Daniel Kahneman, *Thinking, Fast and Slow*. New York: Farrar, Straus and Giroux, 2011, pp. 31–78.

Acknowledgments

In the 15 years since my first book, *Total Project Control: A Manager's Guide to Integrated Project Planning, Measuring, and Tracking*, was published, many people have supported the new techniques that it introduced. Many have applied them in their businesses and further advanced many of the ideas.

Critical path drag computation in a large and complex network can be time consuming, but it is exactly the sort of problem that the computer is designed to solve. Russ Iuliano of Sumatra Development, Inc., Vladimir Liberzon of Spider Project, and Bernard Ertl of InterPlan Systems have included critical path drag calculation in their respective project management software packages. This represents a huge step forward, and I know the day will arrive when no company would dream of marketing project management software that did not compute drag as part of the standard CPM metrics. The other investment techniques will follow.

Over the years, Joe Sopko, Jeff Parker, and Dr. Priscilla Glidden have all introduced the new techniques within their companies. They, along with Denise Guerin, have also been teaching these techniques now for many years in their graduate and executive courses. I sincerely thank them for their missionary work.

Dr. Tomoichi Sato of Japan Gas Company has advanced the techniques of value estimation in a whole new direction with his work and publications on risk-based project value (RPV) analysis. I thank him sincerely both for this and for the graduate classes he teaches at Tokyo University where he has incorporated many of my ideas into his lectures.

Leah Zimmerman has been responsible for expanding and moving some of my ideas into the field of cost engineering. I predict that her creativity and efforts will provide fertile soil for still newer ideas in this field.

Over the years, there have been many companies that have enlisted my services in training and consulting. I trust that they have all found value in these new techniques. But I would particularly like to mention the great pleasure I have received over the dozen years that I have been teaching and consulting with the folks at BAE Systems. There have been too many people to name everyone. But I would especially like to mention Tom Arsenault, Ray Brousseau, John Bugeau, Cheryl Chaput,

Drew Conti, John Dilger, Jim Fasoli, Mark Getty, John Gill, Mike Greene, Bob Korkuc, John Labrosse, Dan Murray, Melinda Norcross, Frank Phillips, Ralph Titone, and Greg Zito. Your organization was one of the first to recognize the value in the new techniques, and I have greatly enjoyed the working relationship we have shared over the years.

I am most grateful to those who have provided me with platforms in academia to teach the next generation of project managers. These include Karen von Sneidern of the University of Massachusetts/Lowell, Dr. Hans Thamhain of Bentley University, Dr. Ken Hung of Suffolk University, Tom Carter and Sybil Smith of Brandeis University, Dave Barrett and Andrew Bennett of Olin College of Engineering, and of course once again, Dr. Priscilla Glidden of the University of West Indies at Cave Hill, Barbados.

Although I have always loved teaching in any setting, my academic students have given me the greatest sense of fulfillment. All the names would be far too numerous to mention. But some that stand out in memory include Ed Anderson, Mahesh Hegade, Dr. Timothy Hemesath, Takayuki Iida, Meena Jayaraman, Emily Ramey, and Vernon Valero of Brandeis University; Jane Maine, Alison Sullivan, and Rich Takvorian of UMass/Lowell; Hang Bui, Vasudevan Devarajan, Jason Edmunds, Anthony Giuffrida, Andrew Masnyj, and Sukhpreet Rana of Suffolk University; and especially my students from the University of the West Indies: Peter Alleyne, Neil Broome, Sharon Carter-Burke, Sidney Cox, Octavia Gibson, Rey Moe, Adrian Sinkler, and Calvin Watson. I have learned so much from those I have taught.

I have always been aware of the value that these techniques offer for those endeavors where human life is at stake. But it was Dr. Adedeji Badiru and Major LeeAnn Racz of the Air Force Institute of Technology who recognized its importance and invited me to contribute a chapter on computing critical path drag and drag cost in emergency response planning to their invaluable publication from CRC Press, *Handbook of Emergency Response*. I sincerely thank them for helping me to spread these techniques into this most crucial area.

The folks at CRC Press have been wonderfully supportive, and I thank Cindy Carelli, Joselyn Banks-Kyle, and Michele Smith for all their help.

Finally, I would like to thank the person who has taught me more about the world than anyone else: my wife Deb, without whom life itself would be impossible.

About the author

Stephen A. Devaux, PMP, MSPM, is president of Analytic Project Management (APM), a training and consulting company he founded in 1992. APM is a Global R.E.P. of the Project Management Institute (PMI). Their clients include BAE Systems, Siemens, Wells Fargo, Texas Instruments, Wyeth Pharmaceuticals, iRobot, L-3 Communications, American Power Conversion, Irving Oil, and Respironics.

Devaux is the author of the book, *Total Project Control: A Manager's Guide to Integrated Project Planning, Measuring, and Tracking* (1999, Wiley). He has worked to develop and use new approaches and metrics in project management with clients in a wide range of industries. "When the DIPP Dips" was published in the *Project Management Journal* in 1992 (an article that was reprinted in PMI's *Essentials of Project Control* in 1999). He has contributed chapters on his new scheduling metric, critical path drag, in two 2013 books: *Project Management in the Oil and Gas Industries* and *Handbook of Emergency Response*. He has authored numerous articles and PMI webinars, and is a frequent speaker at PMI chapter meetings throughout the United States.

He began his career at Fidelity Investments, Citicorp, and the Federal Reserve Bank of Boston and then taught and consulted in project management at Project Software and Development, Inc. (PSDI). He has taught graduate project management courses at Suffolk University, Brandeis University, and The University of the West Indies/Barbados and in executive education programs at Bentley University and the University of Massachusetts/Lowell.

Introduction

The glossary of the fifth edition of the *PMBOK® Guide* defines a project as "A temporary endeavor to create a unique product, service or outcome." The field of project management is then based on figuring out: what kind of endeavor is it?

If we were to define ice hockey to a visiting Martian as "an activity that takes place on frozen water, with hooked sticks and a round disk of rubber," would we expect the Martian to understand what ice hockey really is? It would be crucial to explain what kind of "activity," that ice hockey is a game, a contest, a team sport. Why does someone want to watch ice hockey? Why does someone want to play ice hockey? What is the value of ice hockey?

Defining precisely what kind of "endeavor" a project is leads to being able to think about the value for a project and utilize appropriate techniques for managing a particular type of effort. Every project is an investment, yet the tools that we use to plan, manage, and track projects are different from those used on all other investments.

Merriam-Webster's On-line Dictionary defines *investment* as "the outlay of money usually for income or profit: capital outlay; also: the sum invested or the property purchased." Surely this provides a perfect addition to the definition of a project:

> An investment in work to create a product, service or outcome that is expected to have greater value than the capital outlay.

Every other investment one can think of—stocks, bonds, real estate, anything—starts with consideration of expected benefit: return-on-investment (ROI), profit, net present value (NPV), or expected monetary value (EMV), the term we use in this book. If a project is an investment, then this value aspect of the project, by whatever name, should be its prime operating metric: quantified, planned, tracked, and optimized through every management decision.

Yet, unlike with every other type of investment, a project's expected monetary value is not viewed as a management metric: most project management software does not even permit input of such data. The rare project management software packages that allow such input invariably fail to perform adequate analysis and tracking of the data, analysis and tracking that is crucial input for project decisions.

As mentioned in the preface, the current edition of the *PMBOK Guide* includes a new section that is arguably the most important new section in the entire publication. It describes what it terms *business value*: "(T)hrough the effective use of portfolio, program, and project management, organizations will possess the ability to employ reliable, established processes to meet strategic objectives and obtain greater business value from their project investments." The text goes on to say: "While not all organizations are business driven, all organizations conduct business-related activities. Whether an organization is a government agency or a nonprofit organization, all organizations focus on attaining business value for their activities. " In fact, this business value is the entire purpose of every project and every program in every corporation, government agency, or nonprofit! Recognition of this is crucial.

The open question is precisely how to plan and measure and track and maximize this business value. That's what this book explores. It focuses on traditional and standard project management techniques, such as critical path scheduling, resource leveling, and, of course, earned value, but always from the viewpoint of how to use them to increase business value (which I continue to call expected monetary value to emphasize that it is subject to risk). All of the above must be viewed through the new lens and in the language of investment management. We introduce the reader to new techniques and metrics that guide risk and opportunity assessment and lead both project manager and organizational decision making on the basis of maximized value generation per invested dollar.

There is an important point here that we must make: although practically all programs and projects are investments that require money to obtain and support the necessary resources to do the work, we are well aware that not all value is easily measured in monetary units. For example, the Centers for Disease Control and Prevention (CDC) does not fund the development and production of a flu vaccine in hopes that they will be able to sell it at a profit to the public. Instead, the value of the CDC's investment will come in the form of what are sometimes termed "intangibles": fewer deaths, less suffering, and fewer days of lost labor productivity.

There are many projects like this, where considerations of strictly monetary return may seem trivial or even insulting. In the 2013 book by Dr. Adedeji Badiru and Major LeeAnn Racz titled *Handbook of Emergency Response: A Human Factors and Systems Engineering Approach* (CRC Press), I wrote Chapter 21, "Time Is a Murderer," about critical path scheduling following a major natural disaster. On such a project, it is crucial to recognize that hundreds of human lives may be hanging in the balance as time ticks by. The resources needed to save time may be measured in dollars, but the cost of time must be measured in the human lives at stake. If estimating each life in monetary units (estimated by various US government agencies as being between $6 million and $9.1 million each) helps justify the time necessary to plan, equip, and practice rapid response, then such a monetization may be worthwhile. But what is all-important is the estimate of the cost in lives lost in each different hour (growing numbers as trapped and injured victims start to die), so that the emergency response timetable can be optimized to save the maximum number of lives.

Where it is difficult to translate the value(s) to be generated into monetary units, making smart "investment-based" project decisions usually requires analysis and estimation. In the flu vaccine example, spending an additional $100,000 on the correct critical path activity might allow the CDC to start distributing flu vaccines 3 days earlier. Is it worth the additional investment? Unless there is detailed analysis (with or without monetization) of the amount of death, suffering, and sick days that would be avoided by finishing 3 days earlier, how can anyone ever judge whether the project team should spend that extra money? Such decisions are made every day; it would be nice if they were made on the basis of rigorous cost–benefit analysis.

Such "business" decisions should not be the responsibility of the project team, but rather of the customer/sponsor, whether governmental or corporate. However, looking for such opportunities is exactly what team members are qualified to do, as they are the subject matter experts who can identify how to compress the schedule by 3 days for $100,000. This is precisely the sort of value-added behavior in which the project's customer/sponsor/organization should be pressing the project team to engage. Unfortunately, not only is it not being done, but several aspects of the way we define, measure, and engage in projects can actively discourage such behavior, and these often have negative impacts on business value generation.

In other disciplines, we do not judge investments simply as successes or failures. We recognize that their results are relative, and that more value return is better and less is worse. This is the approach we take, moving beyond the ossifying fixations with "deadline" and "budget" and the binary qualifications of projects as "successes" or "failures." Through the consideration of every project and program as an investment, all members of each organization will be provided with the metrics and tools that allow decision making on the basis of greater or less return, thus letting each associate identify those opportunities that will maximize his or her contribution to the bottom line in a quantifiable way.

Ultimately, this change in approach, to performing and evaluating projects based on business value, is a radical departure from the traditional attitude: one of delivering the product as defined in the requirements on a certain date for a certain cost. Ask any project manager who's been in the business for a few years and she will tell you that this defines her job. "Why should I look for opportunities to increase business value through added scope, better quality, or faster or cheaper completion? That's not my department! If the sponsor/customer wants the project done differently, he should've said so." Even in situations where a simple change could obviously add value, the project team will rarely even suggest such a modification. Stick to the knitting and don't get fancy, is the attitude. And the result is that vast amounts of potential business value are ignored on almost every project.

One can hardly blame the project manager or team. Completing the sponsor/customer's requirements on time and on budget is often a difficult goal, and requires a stressful effort. In addition, both team members and project manager are typically assigned to the project for their technical skills. They may have been given some training on leadership or negotiation, and some of the fundamental techniques of project management, but that training almost never extends to financial considerations more complex than the hourly cost of a resource.

With the acceptance that projects are investments undertaken for their business value must come recognition that this factor is the central consideration for all projects. And this requires at least an awareness of the economic aspects of a project. It is by formalizing the investment arithmetic into project metrics and documentation and reports, data to which the team has to pay attention, that considerations of the project business value will become a part of the day-to-day culture. And then improvements will flow because of what is being measured.

chapter one

Redefining projects

"Why spend money if not to make money?"

You have just been made manager of a project that has a budget of $5 million. You are told that we are doing this project under a fixed price contract. The contract calls for our company to be paid $4 million upon delivery.

What's the matter? Why are you looking at me like that? You have a question, you say?

If you don't, then you certainly should. Who in their right mind would ever pay more for a project than the value they expect to get from it? A sophisticated project manager would immediately ask what's going on. Is there some value other than the payment from this particular customer that would make the project worthy of investment?

Value drivers

There are many, many other considerations that could cause an organization to implement a project whose initial revenue might be less than the cost of the project. Perhaps if we make this customer happy, he will give us more, and more valuable, projects. Or perhaps the technology we develop for this customer can be developed into a product that we can sell to other customers. Or perhaps after the customer receives this product he will need to purchase other products or services in order to keep the initial product operating efficiently. Maybe there is what's called an availability premium at work here: if we don't accept this contract, we won't have any work for three experienced staff members and will have to lay them off, resulting in the loss of their expertise.

The number of possible additional value drivers is large, but not infinite. A list of some of the more common ones can be found in Figure 1.1, where the table sorts some common value drivers of projects into those that add value through *product scope* (i.e., the components, features, and requirements of the final product, service, and result) and *project scope*, also sometimes called *work scope* (i.e., the work that creates the product scope and ensures that the requirements are met). There may sometimes be value drivers that are not included in the table, but it would be extremely rare for them to be adding significant value. This table, which I have utilized for years on consulting assignments, provides a very

Scope Value Drivers	
Product Scope	Project Scope
1. Customer satisfaction	1. Customer satisfaction
2. Revenues	2. Market history/reputation
3. Follow-on (production) revenues	3. Community goodwill
4. Patents	4. Project performance
5. Market visibility	5. Valuable research
6. Product reputation/branding	6. Economies of scale
7. Market dominance	7. Employee morale
8. Savings	8. Staff expertise
9. Product development technologies	9. Resource availability
10. Reusable technologies	10. Customer access/relationships
11. National benefits/security	

Figure 1.1 Some common value drivers of scope.

usable checklist to see what factors may be adding value. Any that are adding significant value on a particular project should be specified as part of the project's business case documentation, along with an estimate of the amount of value.

On projects such as the one described above, where the stated value seems less than the budget of the project, this list may uncover the source of the $1 million plus of value needed to make business sense of the project. And often, when it seems that the necessary additional value driver does not exist, there may be other motives at work, such as an executive sponsor who hopes the project will raise her visibility enough to command a promotion, or a politician who believes the project will increase his popularity by adding jobs in his district.

Some projects are said to be "mandatory," perhaps in order to comply with governmental corporate regulations, for example, installing wet scrubbers in a manufacturing plant to remove pollutants from exhaust and thus comply with clean air regulations. In such projects, the value comes from compliance, which thus avoids either fines or complete shutdown of the facility.

But when all is said and done, we always know that, barring ignorance, insanity, or malice, the total value expected to be generated by any project, from all factors, will always be expected to be greater than the cost. Not that it always works out that way: we all can think of situations in which a project costs much more than anticipated or never produced the value for which the sponsors had hoped. Life does not always go as we anticipate, especially on projects that are, by their very nature, risky endeavors. But we can say with confidence that, at the start of every project, its total expected monetary value (EMV) or net present value (NPV) is forecast to be greater than its cost (i.e., budget).

So when presented with a new project such as the one above that seems like a value loser, we understand that such a scenario is either improbable or incomplete. That is because we intuitively understand that every project is an investment. We recognize that investing $5 million in a project that is only expected to return $4 million in value makes no sense. Instinctively, we understand that there must be other value drivers that add up to more than the $1 million that this project would otherwise be destined to lose.

Planning and tracking projects as investments

The fact that all projects are investments should surely mean that the expected monetary value of every project should be managed, planned, and tracked as with every other type of investment. It also means that if the factors that drive that investment value are hidden from the project manager or team, they cannot be planned properly and optimized for their value contribution. On a project, such optimization should be the job of the project manager and leading team members because they are the people who understand the details of the project work and can generate and analyze ideas that can lead to greater efficiencies. But they need key items of project information in order to be able to contribute in such ways and, unfortunately, all too often such information is not provided by the sponsor/customer. The mandate is almost invariably simplistic: provide X product on Y date for Z cost.

In addition, the ongoing progress of the project in terms of the value-above-cost that it's expected to generate should be the most important project management metric to be tracked during project execution. Yet, because recognition of projects as investments has barely begun, such tracking is routinely omitted. Instead, projects are tracked on the basis of three items:

1. How is the project doing in relation to the planned schedule?
2. How is the project doing in relation to the planned cost?
3. And (often almost as an afterthought) how is the project doing in terms of the requirements and technical goals of the product scope?

The reason that the third item is often an afterthought is due to the fact that it is unquantified: schedule and cost data are both measurable and comparable to the project plan in mathematical ways, allowing for subtraction to show differences and percentages to show trends. These are the essence of earned value (EV) metrics such as schedule variance (SV), schedule performance index (SPI), cost variance (CV), and cost performance index (CPI). These metrics are, or can be, very useful tools, and we cover them, the good, the bad, and the ugly, in great depth in Chapters 8 and 9. But notice that none of these metrics speaks to the *raison d'être* of the project, the reason we are investing so much time and money: to generate the project's product, service, or result.

Traditional project tracking

Let's consider the implications of this omission in terms of judging project performance. Take the following scenarios, each judged in traditional project management (PM) terms, and all completed. Assume that each project had a $5M budget, a 12-month deadline, and similar or identical scope. Which was a better result? Which was worse?

- Project A costs $7M, but finishes in 12 months with 100% of the requirements met.
- Project B costs $5M, but finishes in 14 months with 100% of the requirements met.
- Project C costs $5M, and finishes in 12 months, but with 80% of the requirements met.
- Project D is canceled after 3 months, with $2M spent and no requirements achieved.

Without investment data, specifically the expected monetary value that is detailed to specific items of scope and the value/cost of time, as far as we know any of the above could be the best result. Furthermore, these are just four possibilities of an infinite range of possible results. There are others:

- Project E costs $4M, but finishes in 11 months with 70% of the requirements met.
- Project F costs $8M, but finishes in 10 months with 105% of the requirements met.
- And so on.

Remember, each of these projects is finished. Each of these results was achieved not by accident, but by management and decision making. How on earth can project teams make decisions, or even look for opportunities to improve value, if they will not know, even after the project has been completed, what was the best result? Instead, with the current metrics, teams are simply charged with marching blindly to a budget/deadline cadence (and one that may be completely unreasonable and unachievable).

Project success and failure

Completed projects are routinely separated into a binary division of "success" and "failure." What do these terms mean? Is a project a success or a failure if it was (a) completed by its original deadline, and (b) within budget, but (c) the functionality of the final product was reduced by 30%? Is it a success or failure if the product functions exactly as conceived, but it took 20% longer and cost 40% more than planned? Surely that determination must depend in large part on the quality of the original estimate that was used to generate the "deadline" and "budget."

How are those schedule and cost parameters known as deadlines and budgets determined? One would like to think that a great deal of analysis goes into setting such boundaries; for make no mistake, once the deadline and budget have been set, they will acquire an autonomous power to influence both the results of the project and the behaviors of personnel in the organization, both within the project team and beyond. This influence may sometimes be beneficial, but it also can be malignant. The fact that the requirements of the final product are often hard to quantify makes that aspect of the project particularly subject to invisible pruning when the more quantitative aspects of the project (i.e., time and cost) are at stake. This is the primary reason why quality is often sacrificed on projects, and why systematic quality checks must be mandatory on projects. The 2013 rollout of the US government's Healthcare.gov website is a prime example of this. It would have been much better to delay the rollout until adequate testing and remediation could have been performed rather than delivering an untested product on schedule and hoping for the best.

Of deadlines and budgets

So how are those deadlines and budgets derived? Does the sponsor/ customer:

1. Go back and look at a database of 10 or 20 previous projects that involve similar work?
2. Determine precisely which aspects of those projects are similar to this new project?
3. Analyze and average the duration and cost actuals from the previous projects' work packages?
4. Add or subtract a little more for work that seems a little harder or easier?
5. Build duration and cost parameters that seem reasonable?

Sure, that is done sometimes, and organizations that conscientiously collect project data on actual cost and time at a reasonably detailed level, and assemble work breakdown structure (WBS) templates based on those actuals, do exist. Such companies usually perform a pretty good job of creating what's called a *basis of estimates* (BOE). But, sad to say, such maturity in project management methods is by far the exception, limited to the most mature PM disciplines such as large construction projects and US Department of Defense (DoD) contracting. And even in such industries, the data selected as "similar" are often infected by wishful thinking.

And more often than not, deadlines and budgets are plucked from thin air, with little or no methodology more complex than what can be termed "whim."

Procrustean bed of deadlines

> Add an *s* to "saltines" and you get *saltiness*.
> Add an *s* to "timelines" and you get *timeliness*.
> Add an *s* to "deadlines" and you get *deadliness*.

And that indicates what deadlines often do to projects: instead of an active and animated team, striving to create the maximum value from their work in the most efficient way possible, we reduce the effort to a humdrum regularity where everything is hunky-dory until it starts getting close to that deadline. Then the whole organization suddenly starts panicking and running in circles like a headless chicken to figure out ways to avoid going one day beyond. And all this happens no matter how arbitrarily the deadline was originally contrived.

In 1955, Cyril Northcote Parkinson first expressed what would become known as Parkinson's law: "Work expands so as to fill the time available for its completion." Although most people today are not familiar with Mr. Parkinson, his dictum remains as true as it ever was. And not only does work expand to fill available time, on projects it also expands till it exhausts the available budget. A deadline and a budget almost guarantee that the project will not finish ahead of schedule nor under budget.

1. If a PM completes a project, meeting all necessary requirements, on budget and on time, does that make the PM a good PM or does it merely mean that the project was technically easy with an exorbitant budget and a very "loose" deadline that anyone could have achieved?
2. Alternatively, if a PM cuts scope by 20% and still finishes a month late and 15% over budget, does that make the PM a bad PM or does it merely mean that the project was impossible to achieve within an absurdly tight deadline and budget, and the sponsor was unwavering in her demands? Is it not possible that rather than the project being a "failure," it was actually brilliantly managed, and the very best possible result was achieved?

We must start realizing that, if projects are treated as investments, then duration and cost estimates are merely targets intended to yield a certain level of project profit.

The term "deadline" has a very real history and meaning; it was a line set up approximately 20 feet inside the fence in US Civil War prison camps. Any prisoner venturing beyond that line would be shot by the guards. That was a real "deadline." Perhaps because of the military roots (i.e., US Department of Defense) of many project management traditions and techniques, the deadline has become common parlance and even acquired the status of a management and contractual technique. But by analogy, a deadline in project management terms surely should mean the date at which a project dies, that is, loses all vitality and investment value. In those terms, such scenarios are extremely rare in project management. Instead, what usually happens is that projects lose a percentage of their expected value as their duration increases, but rarely does that expected value reach zero even after months of delay unless the project is aborted. It does not mean that knowing how much value is lost for each increase in duration is not vital information. But it does mean that project metrics should reflect the relative nature of an investment instead of the crude and overly dramatic binaries described by terms such as deadline and failure.

Integrated metrics linked to expected project value, and demonstrating the variable impacts of time, cost, and risk, would allow determination of which of those project completion scenarios listed previously would be, in terms of the combined value/cost impact of scope, duration, and resource usage, the best investment outcome. They would also permit the PM and the team to be the sponsor's allies in looking within the details of the project work for opportunities to increase expected project value in the integrated investment metrics set up by the sponsor/customer at initiation.

Quantifying the triple constraint model

Consideration of every project should start with the model depicted in Figure 1.2. The diagram has existed for about 50 years. It is a durable model in project management literature, known as the triple constraint model. It reflects the fact that every project is a compromise among the three sides: SCOPE, TIME, and COST. If this were all the model had to tell us, I would not have bothered including it in this book. It represents

Figure 1.2 The triple constraint model of a project.

the integrated reality of a project; this is where the quest for integrated project metrics must start.

This model can be used to represent the nature of the project as an investment:

1. *COST* on a project is really a measure of resource usage. Different types of resources are measured in different units: liters of paint, kilograms of nails, or hours of labor. But all these various resources cost money, in one form or another, and so the cost of the liter or kilogram or work hour is translated into money. That planned amount of money to be spent on resources is the *budget*.
2. *SCOPE* is, in the *Guide to the Project Management Body of Knowledge* (*PMBOK® Guide*) terms, two different things: *product scope* and *project scope*:
   a. Product scope is the features and requirements of the product, service, or outcome that the project is undertaken to create.
   b. Project scope, sometimes referred to as *work scope*, is the work necessary to create the product scope, the output of the resources that the cost pays for.
3. *TIME* is the elapsed duration from the beginning to the end of the project. Although there are sometimes components that are delivered earlier (subdeliverables), a project ends with the completion or delivery of the total integrated product, service, or result that meets the specified requirements for which the project was undertaken. (Note that a *program*, which we discuss later, is defined differently and therefore has different completion criteria.)

We now explore each of these parameters in terms of how they may interact with each other as investment factors on a given project.

COST side of the project investment triangle

From a project performance viewpoint, it is critical to know certain key information about each resource that the project requires:

1. *What are the precise features of the resource?* Blue paint is different from red paint, 3-inch nails are different from 2-inch, and an electrical engineer with five years of optics experience is different from one just out of college. The project manager needs to specify precisely what is needed. If what is assigned is different from what was requested, that difference should be reflected, not just if there is a change of cost but also in any increased risk to schedule or scope.
2. *How much of the specified resource is needed?* Would an increase in the amount of a resource assigned to a specific activity result in

shortening the duration of that activity? If so, by how much? In the chapter on CPM scheduling, we refer to a concept called the *doubled resource estimated duration (DRED)*, which is a reflection of increased resource usage on activity duration.

3. *When is the specific resource needed?* As we show when we discuss critical path scheduling and resource leveling, this is a key data item in managing a project efficiently. Having a resource assigned to the project at a time when it's not needed increases cost, and not having the resource when it is needed often leads to delays. Are those delays on the critical path? If so, they will increase project duration.

Each of these items of information about resource usage not only will affect the COST side of the triangle, but also may interact with and change the SCOPE or TIME sides.

Budget and cost are of primary relevance to finance and accounting. But because budget is also the amount of money that the sponsor is investing, it is clearly also of great importance to both the sponsor and the project team. The sponsor therefore establishes a budget for the cost of all the project resources. This then is intended to serve as a cap on the amount that the sponsor invests. Any proposals to raise this cap, due either to scope changes or excess resource usage or cost, must be approved by the sponsor.

For the project team, the budget is then decomposed to provide interim metrics, based both on periodic spending (usually weekly or monthly) and on activity-based work packages within the project. This provides the basis for earned value metrics and tracking, which the finance department collects and analyzes, which the project team uses as a spending guide, and which the sponsor/customer checks for early warning signals of trends that might lead to overspending.

Yes, the project team should endeavor to stay within the budgetary constraints. However, it is sad to say that project teams almost invariably operate as though the budget included the purchase of blinders. They see their mandate as being simply to stay "between the lines," within the confines of that budget. (Not much inside it, mind you, a single dollar will do!) And indeed, in order to ensure that they do so, they will if necessary cut corners elsewhere, clandestinely if possible, because going over budget is very easily detected.

It is extremely rare that a project team will seek opportunities to increase spending, potentially exceeding budget in order to increase the expected value of the sponsor/customer's investment. The typical project manager and team usually have no idea of how to do this, nor that looking for ways to do so should be part of their responsibility. And when project managers attempt to do this, they are frequently slapped down by the corporate culture. ("What? You want an extra $20,000 to make this product better? You are already at your budget limit! Just deliver what you are told!")

There are three basic ways by which a project team can, by spending extra money, enhance the value of the project investment:

1. Increase the scope.
2. Accelerate the schedule.
3. Reduce the risk.

These three project changes are what opportunities on a project investment look like. Notice that neither the project manager nor any member of the team has license to go over budget without approval from the sponsor(s). However, if they see (because they're looking!) an opportunity to have a positive impact on the project investment's expected return or profit through a small increase in cost, shouldn't they bring that fact to the investor?

Of course, the sponsor/customer should always have the right to say: "No, I don't have the cash to do this," or "No, I don't think it will improve the investment significantly." Then the original budget remains in place. But surely most investors would like to be made aware of such opportunities. They should attempt to encourage such initiative by providing incentives (promotions, raises, bonuses, or even just plain "Attaboys") to those who increase the value for both the project and organization.

But the standard project metrics that mandate a fixed scope for a fixed budget at a fixed deadline suggest no wriggle room for creativity. And therefore the investor (and the organization) should change the standard metrics in ways that would reflect positively on such discovered opportunities. (Hint: that is where this book is headed.)

Finance departments and overhead burdens

In any investment, the amount of money that will have to be invested is a crucial metric. Let nothing in this book seem to downplay the vital importance of having good planning and tracking data for COST management. However, planning and tracking the cost of resources is one area of project management where most companies have invested heavily and are doing a very good job. There are whole departments of accountants and finance experts and IT personnel, and expensive cost-accounting computer systems, all busily keeping tabs on COST. In fact, in many project-driven organizations, finance departments have more clout over project decisions than the members of the project team.

Unfortunately, most finance people do not understand the specifics of project management techniques, particularly the concept of the critical path. As we show, this lack of knowledge is extremely important, as the mere fact that work is occurring on the critical path changes its implications: a minimum-wage laborer performing work on the critical path can cost the project and the organization more than a Nobel laureate physicist working off the critical path! We discuss this extensively when we get to critical path scheduling and the new concepts of the drag cost and the true cost (TC) of project activities.

One item to which most finance departments pay a great deal of attention is overhead. Indeed, this is very important, as overhead is an element in what an engineering executive of a large aerospace contractor that I used to work with referred to as *marching army costs*. As long as the army is marching, it must be clothed, fed, and sheltered. And all that requires money. So too, as long as the project team is working, it must be provided with office space and heat and electricity and health insurance and paychecks and a variety of other support functions that cost money. No mature organization would ever be so naïve as to believe that the cost of an employee is simply salary; that would be a good way to go out of business in a hurry. So corporate finance departments spend a good deal of time performing analysis as to what the overhead burden is on employee salaries.

Overhead burdens are sensitive to geography: the cost of rent is very different for a manufacturing plant if it is located in Chennai versus Akron versus Manhattan, and other overhead inputs would also vary. Multinational companies track overhead costs for each country and city, and sometimes for multiple physical plants within the same city. However, the consequence of tracking overhead costs, no matter how conscientious and detailed, results in a rather amazing peculiarity: all the employees at a given plant, from assembly line worker to general manager, will have an overhead burden percentage that is exactly the same. Isn't that amazing! Every one of 4,000 low-level workers may have a 50% overhead burden, suggesting that each of the 4,000 uses exactly the same amount of electricity and has exactly the same amount of desk space as all the others. If their salaries range from $500 per week to $1,000 per week, the overhead burdened cost for each employee will range from $750 per week to $1500 per week, not approximately but exactly 50% more than each employee's specific salary. And the general manager whose salary is $4,000 per week? Some companies will make an adjustment for the overhead burden on senior managers. But most don't, meaning that the general manager of the plant also will be assumed to carry an overhead burden of exactly 50% of salary, for a total overhead burdened cost of $6,000 per week.

Amazing, isn't it, how everybody's overhead burden is exactly the same percentage of their salary? Actually, you probably don't think it's amazing, because you are aware that the overhead burden is merely an estimate, approximated by dividing the total cost of running the plant over the number of employees at the location. We are comfortable with this sort of estimation because, even though we know that it is hugely lacking in precision at the individual employee level, (a) it is useful at the macro level and (b) everyone is used to it.

But remember this practice when we talk of estimating other aspects of the project investment, such as expected monetary value. Because, in my experience, when we get to other data where a quantified approximation would be extremely useful, objections of, "But that's just an approximation!" suddenly spring from everywhere.

The truth is that good decisions in all investments, from poker hands to hedge funds, projects, and programs, rely on mathematics and quantified data. If there are no quantified data, decisions will be made anyway, only they will be based on assumptions that have been made up along the way. It is much less risky to do a modicum of due diligence and try to approximate what the real numbers are than to go with a "gut feeling" that may be 180° wrong; some people's guts are a real mess!

SCOPE side of the project investment triangle

More than a quarter century ago, after several years in the financial industry working on project teams and even as a project manager, I finally learned that project management was not simply a matter of applying common sense to work situations. That was the morning that I started a new job with a company that made project management software. With my "strong background" in project work, I was put in charge of developing training courses in the theory and practice of project management. And on the first day on the job, I discovered that project management was not what I and all my previous bosses considered it, a seat-of-the-pants application of common sense to work goals, but rather a specific discipline that included an array of quantitative techniques.

At that time, most of the hundreds of commercially available project management software packages marketed themselves in terms of something that was called *cost/schedule integration*. This was the ability of the software brand to reflect the impact of any change in schedule on the cost, and of any change in cost on the schedule.

But I soon learned that there was a third parameter, variously referred to as *scope, requirements, technical* or *quality*. So I asked some senior people in my company, "What about that third side? Would our software help users see the impacts of changes in scope on the other two sides of the project plan?" And the answer was no: "Scope is a constant," I was told. "But for any constant scope, the software will show the schedule and cost."

The idea that SCOPE is any sort of "constant" is not only easily dismissed, it is also dangerous. Product scope takes work to accomplish; that work takes time and requires resources, which cost money. And therefore it is immediately obvious to any individual employed on a project that it is usually possible to reduce either duration or resource usage by reducing scope.

One of the key elements of SCOPE is quality. Indeed, as pointed out above, it is not unusual to see the triple constraint model displayed with SCOPE replaced by QUALITY. This is an error as factors other than quality can affect TIME and COST: product features and sales/marketing, for example. In fact, quality is simply an important component of SCOPE, and quality checks such as component and integration testing are methods of ensuring that the SCOPE meets the stipulated requirements.

As mentioned above, the *PMBOK Guide* defines scope, quite correctly and importantly, as two different things: *product scope* and *project scope*. Let us deal first with product scope.

Product scope: Main generator of project value

Product scope is the features and requirements of the product, service, or result that the project is undertaken to create. It is the most important aspect of a project. In my seminars, I will often ask the attendees which of the three sides of the triple constraint triangle is most important, and I will usually get the answer: "It depends." These two words provide a good answer to many questions in life, however, not to this particular one. SCOPE is the *raison d'être* of every project: its reason for existence. Even in cases (and they do and should exist!) where we decide to do a project simply because we need to keep our skilled staff employed and thus available for future projects (a form of availability premium), we still assign them to projects where the product scope is chosen as being the most valuable and suitable. On any project, although we may be willing to compromise on the requirements for reasons of TIME and COST, the eventual product, service, or result is chosen because of its value. And that value is the reason for the investment of resources, money, and time. In other words, it is the motivation for the investment of COST.

On most projects, the product scope generates all or the vast majority of the value for which the project is undertaken:

- *The revenues that a new product will generate.* Here it is important to remember that the new product will at some point become an old product. Obsolescence is out there in the future, waiting to sink even the snazziest new technology beneath the ocean of history. We may not know what that date will be, but it is important to estimate it as it will have an impact on the product's value.
- *The follow-on revenues that a new product will enable* through either additional products or services that would be impossible without the initial *enabler project* (see the section on projects within a program). In any program of related projects, the value of one

project can often kindle the value of other projects. For example, the toys and other merchandise that are produced and marketed based on a hit movie can often kindle DVD and on-demand sales of the movie itself. Each effort can be viewed as a separate project, with its own separate value. But if all the products (film, toys, DVDs, and on-demand sales) and the projects to market them are viewed as related aspects of a single program, better decisions can often be made to maximize the total revenues. And if we do not see the value of those future projects as being attributable to the enabler project, we may not only underestimate the expected monetary value of the initial project (in this case, the production and release of the movie), but also make bad project management decisions regarding scheduling and resourcing. To give a simple example, imagine that we are making a movie about Softy the Snowboy, which we initially plan to open in the theaters in late October. However, we discover that we might somehow be able to save some money by delaying the production of the movie by a month. There will be no negative impact on ticket sales as they would actually be maximized by scheduling the opening in movie theaters for the day after Thanksgiving instead of a month earlier, when the popularity and attendant buzz might fizzle away before families start flocking to the big shopping mall megaplexes. Thus the projected profit of just the movie project would seem to be enhanced. But what if this means that sales of toys, videos, and music based on the movie won't reach their peak until a lot later, so that we miss much of the holiday shopping season and actually decrease the value of the Softy merchandise project? The only way to visualize this is to recognize that the value of the program enabler project is the sum of both its ticket revenues and the kindled impacts of the follow-on projects. (We discuss this further in Chapter 2 when we consider the impact of time on the value of enabler projects.)

- *Savings that will be generated for the organization* through a project to create a product or service that will lead to greater productivity or other efficiency. Projects are not always about generating money; sometimes they are about saving money.
- *Expanded market share due to a new product* that, although it may be limited in its revenue generation, increases the organization's profile in a specific market space. Again, a portion of the value that an increase in future market share will add may be attributable to this initial product, and we should take that into account both when we select the initial investment and as we manage the initial project. It is part of the equity that drives the initial project investment.
- *Patents for components that can be used on other products* that will provide market exclusivity or that will generate revenues from other

companies through fees. Of course, patents are also time-limited, and, as with obsolescence, this must be taken into account in the management of the project resources and schedule.

- *Public good/national security, including the saving of lives, avoidance of injury or illness, and protection of property.* If marketing and finance departments think they have a hard time estimating the dollar value of a project motivated by revenue or savings, pity those folks who work for governmental and quasi-governmental agencies. Organizations such as the US Department of Defense (DoD), the Department of Homeland Security, the Federal Emergency Management Agency (FEMA), the Centers for Disease Control and Prevention (CDC), the Occupational Safety and Health Administration (OSHA), the Environmental Protection Agency (EPA), the Food and Drug Administration (FDA), or any of the thousands of similar organizations throughout the world rarely see themselves as making investments. But that is exactly what they do, using taxpayers' money. Surely they have a duty to ensure that they are spending public funds in ways that will most benefit the populace? Within some organizations, such cost/benefit analysis is being performed primarily to make sure that projects are not being implemented where the benefit will be far less than the budget.

In 2011, the *New York Times* reported that the EPA valued human life at $9.1 million, the FDA at $7.9 million, the US Transportation Department at around $6 million, and the Department of Homeland Security was suggesting that preventing deaths from terrorism should be valued about 100% more than from other causes, presumably due to the potential collateral effects of fear and economic suffocation.[1] Different methods have placed the value of a lost year of human life at anywhere from $50,000 to $129,000.[2] One can certainly haggle over the numbers, but the sort of cost–benefit analysis needed for decision making requires a foundational quantification. Projects that save lives need resources, and resources are limited and cost money. Most people would like to think that $20 million is not being spent to save a single life if there is an alternative to spend that money in a way that would save 20 lives.

SCOPE is of primary relevance to the customer(s) or the sponsor(s). The value of the scope, both product and project, is the reason that customers and sponsors are willing to invest the money for the resources to do the work. It should therefore be standard procedure on all projects for the sponsor to ensure that adequate cost–benefit analysis of that scope is performed before the rest of the project is planned. The various aspects of that information can be hugely useful to the planners: their decisions will affect the value of the scope and so should be provided as guiding metrics. This includes how quality, time, and risk may affect the value of the scope.

However, as we show in the next chapter when we explore the concept of TIME and its tradeoffs with SCOPE and COST, the decisions become both trickier and more momentous: how many lives a project will save may be just as influenced by time as how much money our remote-controlled Softy the Snowboy or beach hotel or antidepressant pill will generate.

Project scope: A secondary generator of the project value

Project scope, sometimes referred to as work scope, is the work necessary to create the product scope. Project scope is driven by the details of product scope: it is whatever is required to ensure that the final product's intended features, requirements, and quality are achieved. In turn, project scope drives the project duration and the cost of the resources to do the work.

As shown back in Figure 1.1, sometimes the project scope can add to the investment value of the project. For example, there are cases where the resources on staff will learn a new skill through their work with a new technology on a given project, a skill that will have value to the organization on future contracts or projects. In addition, the way that the project team performs the project, such as by keeping in close communication with the customers and making them feel "warm and fuzzy" about the way the project is being managed can lead to future contracts. In these ways, project scope can generate a small but sometimes significant percentage of the project's expected value. Certainly any aspects of the project scope likely to lead to significant follow-on value to the organization should be spelled out in the project's initiation documentation such as the project charter and business case.

If these sorts of add-on values through project scope performance are not itemized and communicated to the project team, significant generators of value in the project investment can be lost. For example, senior management may decide to submit a lower bid on a contract specifically because it will provide our organization with an opportunity to practice a new technology we believe will become standard in another year or two. Even though we could perform the project using our traditional skill set, this new project seems to provide an opportunity for subsidized on-the-job training.

However, no one has taken the trouble to explain this purpose to all the engineers who will actually be doing the work. And, under the pressures of deadline and budgetary constraints, the engineers conclude that it would be faster and cheaper to use the old technology rather than to try to get up to speed with the new. They are quite proud when they bring the project in on time and within budget. But a large chunk of the project's expected value—the important new skills that represented a primary reason for our original bid—have been dropped by the wayside. We would have been better off spending an extra $100,000 wrestling with and learning the new technology because senior management had estimated those skills will be worth $5 million to the organization on future bids over the next five years. But that sort of fact often gets lost in the fog of work.

Customer value and internal value

Whenever a project is performed by a contractor for a customer, important complexities enter the picture and they must be managed appropriately. Suddenly, there are two organizations embarking on the investment instead of one. And recent history with Wall Street investment houses has demonstrated that such situations can be fraught with conflicts of interest and moral hazard. Leading up to the 2007 Wall Street crisis, many financial institutions were serving the role as both investment advisor and purveyor of investments. Clients were persuaded to purchase investment instruments that the broker needed to sell. In such cases, it is very difficult to envision win–win conclusions. There may be great wisdom in the concept of *caveat emptor* (let the buyer beware!), but it does not generally lead either to trusting and symbiotic relationships nor to efficient investments.

The economic risks involved in customer–contractor relationships has been analyzed for over a century under the heading of the *principal–agent problem*. The agent organization, in this case the contractor, is closer to the work and is far more likely to be able to discern opportunities or risks and alternative ways of doing the work than is the customer organization. But insofar as the contract does not adequately align the customer's (principal's) values and benefits with those of the contractor, the customer is the party that is likely to be shortchanged.

This problem makes it absolutely crucial that the definition of every project as an investment be clearly understood, and that the essence of each specific project investment be recognized and emphasized in the terms of the contract. Let me emphasize this:

If the terms of the contract simply enforce achieving the deadline and budgetary constraints, then those will be the primary clauses to which the contractor pays attention, and methods by which the customer could achieve greater return on his investment will be ignored. And if the deadline/budget constraints are not written into the contract in a way that sufficiently affects the agent organization, then even those parameters may be ignored!

The approach of many of those professionals whose job it is to draw up contracts makes the situation even worse. Are there contracts professionals who strive to create agreements that maximize the satisfaction of both sides? Yes, but most such professionals work in fields where customer and contractor are likely to have an ongoing relationship that extends far beyond the completion of the project. Much of the time, the specific contractor and the specific customer on a project may never see each other again in their careers. Such cases are like housing sales, where each side has only one goal: to get the best deal possible, whether that means receiving the highest price or paying the least amount. Indeed, many contracts personnel treat the process as a game, motivated not just to win, but to win by the biggest possible margin. The fact that a short-term win may lead to a long-term loss is usually ignored.

In some ways, one cannot but feel sympathy for the contractors: the responsibility for delivering what may be a very tricky and technologically difficult product is sitting on their desks. And the project manager knows that every day of extra time that can be negotiated until the contractual deadline, and every extra dollar in the budget, will provide a buffer between her and the stomach ulcer medications. If the customer organization truly wants to maximize return on investment, then it must ensure that the terms of the contract do precisely that.

But they usually don't. Why?

1. First of all, because the major thrust of this book, that every project is an investment, is not adequately understood and its implications are ignored.
2. If a project is an investment, then the terms of that investment must be plannable, trackable, and quantifiable as with every other investment. But often neither project personnel nor senior management know how to do this.
3. If the customer is going to turn his investment over to an "agent" to manage for him, he must at least take responsibility for ensuring that the agent has both the knowledge and the incentive to maximize the customer's return. But, for the most part, the customer does not know how to do this.
4. The customer usually relies on professional contracts people, some of whom are probably attorneys. They have studied law, a voluminous and constantly changing subject. But they usually have minimal knowledge of the specific requirements and techniques of project management beyond cost management (i.e., earned value) and therefore only know how to do what they've always done: set deadlines and budgets.

It has become commonplace for me, working with contractor organizations, to hear horror stories of ignorant customer contracts personnel. One common complaint relates to the unwillingness of the customer to allow *management reserve*, either schedule or cost, to be included in the contractor's project plan. "Take that out!" the contractor will say, thus clearly demonstrating total ignorance of project management techniques. Management reserve is a necessary part of any project plan, allowing the project manager to mitigate risk and react flexibly and efficiently when circumstances change. The absence of management reserve simply increases the pressure on the project team. It makes the team far more likely to react incorrectly to unexpected obstacles by taking unwarranted risks in trying to recover to the original parameters. Have you ever found yourself running late for an important appointment, decided that a certain back road might be a shortcut, and found yourself totally lost and horrifically late? That's what happens to teams that don't have management reserve.

Implications of various contract types

During one of the presidential debates leading up to the 2008 election in the United States, John McCain stated that, if elected, he would do everything in his power to require that all Department of Defense contracts be *fixed price* contracts. I smiled to myself and wryly thought, "Your lips to God's ear, Senator."

Yes, it would be nice in some ways if that could happen. On a fixed price contract, the contracting organization shoulders the risk. It is committed to delivering the product or service for a specific amount of money. If the cost turns out to be greater, the contractor takes the loss. As taxpayers, we would rather have cost overruns paid for out of corporate revenues than out of our tax revenues.

However, there is no such thing as a free lunch. What this approach would mean is that, on any defense program designed to develop new technology (and therefore inherently risky), either not a single contractor would bid on it or, if some did, the bids would be exorbitantly inflated in order to cover the contractors' risk. And the result, in all probability, would be gigantic overpayments for the taxpayer.

There is another technique that used to be very popular with contractors, one sometimes referred to as "making up the difference in engineering change orders (ECOs)." In this approach, contractors knowingly underbid their best estimate of the cost of a program in hopes of being awarded the contract as the low bidder. Then the effort would be to understate, as much as possible, the product scope that the contractor was obligated to provide. And every time the customer insisted that he needed more than the contractor was planning to deliver, an exorbitantly overpriced ECO would have to be issued and added to the contract. ("Oh, you wanted toilet seats on those commodes? Well, why didn't you say so? That'll be $895 each.") Fortunately, government auditors have gotten a lot better at squashing the more blatant of these practices.

I know there is a tendency for many people to feel that the sorts of contractual situations that I described above would occur primarily on government contracts. In my experience, this is simply not true; contractors in the private sector are equally likely to have this business reality.

This was brought home to me most emphatically several years ago when I was teaching the first day of a four-day project management class for a company that sold and customized software systems for the insurance industry. After a couple of hours, one of the attendees interrupted me and said, "You keep telling us how to make projects shorter and less expensive. I don't want my system customization projects to be shorter and cost less. I want them to be longer and to cost more."

I nodded: "I see. I assume that means that a lot of your projects are being performed under cost plus terms."

"Almost all of them," he replied. "And when the project ends, not only does the revenue stream stop, but I have to find other jobs to put my people on. The longer I can keep the project going, the more money we make and the more people we can keep busy."

"I see. But don't you wind up with very dissatisfied customers?"

He shrugged. "By the time they get dissatisfied, it's too late. They've paid a few hundred thousand dollars for the system and we've told them that the customization might cost another hundred thousand. By the time they discover that it's likely to cost close to $500,000, they're already into us for so much that they can't quit then."

This is a business reality and an unintended consequence of this type of contract. I was almost lost for words; making projects shorter, more efficient, and less expensive is not obvious and is not incentivized by some current systems.

It is a cliché to say it, but behavior that is incentivized and paid for is the behavior we will get. But whatever else they may be, both corporations and people are economic beings, and as such are likely to behave in a manner that is to their own economic interest. That means that contracts must be written in such a way as to ensure that results which benefit the principal/customer also benefit the agent/contractor. And that means incorporating project-literate clauses that provide incentives to the contractor for taking actions, making decisions, and generating results that benefit the customer.

The best way to achieve this is by abandoning the old techniques of setting deadlines and budgets, and instead incentivizing the contractor to deliver the project in a way that maximizes the customer's investment. Of course, in order to do this, we first have to start recognizing that projects are investments and using metrics that reflect that reality.

Project investment metrics

Ask any project manager how big his project is and you will hear something like: "Well, it's a 10-month project with a budget of $2 million."

Great! But why are you doing it? Because you have nothing better to do with your time and resources? Because you need to get rid of all that money?

The truth is that the most important part of any project is usually ignored: why are we doing it? We are doing it because of the value we expect to get from it. And we are going to get the bulk of that value from the product scope that is delivered at the project's completion.

The implications of failing to use metrics that focus on this are huge. If projects just cost money and occupy time, then who would ever want to do them? As in so many matters in life, it is important to focus on the positive aspects rather than the negative. The focus on project cost makes all projects "cost centers," and everyone knows that organizational cost centers have a much harder time getting resources than "profit centers." And if projects are cost centers, what does that make project managers? It makes them overhead on cost centers! No wonder project managers complain that they are unappreciated.

The truth is that every project is a profit center, selected and performed only if the value of scope is expected to be greater than the cost of the investment. And that difference between the value of the project scope and the budget that must be invested is the prime metric for all investments, and thus for all projects: *project profit*.

Of course, when the project starts we do not know what the value is going to be. Indeed, when the project is finished, we may still not know what its value is. It may take several years and much accounting and analysis after the project's product has been deployed to determine what its value actually has been. This is analysis that, although it is often simply never done (or if it's done, little attention is ever paid to it), can provide crucial data for future project selection. Was the project originally estimated to contribute value, from all sources and drivers, of $50 million? Did it only contribute value of $40,000, or did it do much better than expected and contribute $100 million? What precisely caused either the original miscalculation or the later reduction/increase?

A project is a risky investment, but the risks need to be analyzed up front in determining the ultimate expected value. That process must start by documenting expected value generation at project initiation, and then revisiting the issue periodically. Value generation analysis may need to be conducted many months after the project has been completed.

If a product performs poorly and generates nowhere near the value that was predicted at initiation, is it due to the fact that the sponsoring executive, or even a certain financial analyst, overestimated its value? Does this individual have a pattern of doing this? If so, steps need to be taken to remedy that situation; investment managers and market analysts need to be realistic, not overly optimistic.

Or perhaps the project's completion was targeted for the summer season beginning June 1, but it wound up not being completed until September 1? If so, was that the fault of the project manager or project team? Or was it the fault of an unrealistic launch target date, perhaps set by the same executive? Was it understood up front that a three-month delay would reduce the value by $10 million, and was the risk of such a delay realistically analyzed?

If the fault is determined to be that of a project manager who has had similar problems in the past, we can guess that his future career will be and should be with a different organization. But the same thing should be true of senior managers who overvalue projects that they sponsor, thus expending resources less than optimally. And those who set target completion dates that are unachievable and thus lead to substantial reduction of value (as well as having a deleterious impact on staff morale) are no better.

We must use project metrics that reflect investment information integrating all three sides of the triple constraint paradigm. And the defining metric for any investment is profit: the difference between the money invested and the value of the return. COST measures the amount invested and SCOPE is what generates the return. If on every project we estimate the value of the SCOPE, then the expected project profit will be the difference between those two numbers. And that should be the guiding metric for the project and the project team.

Expected project profit planning formula

$Expected Project Profit (EPP) = $Expected Monetary Value (EMV of scope) − $Budget

Profit should be the prime metric for projects, as it is with all other investments. At the start of the project, this formula should be the determining factor in whether we invest those budgetary funds in this project. When the project is over, the formula will have changed:

Expected project profit at completion formula

$EPP = $EMV − $Actual Cost

The formula has changed, but perhaps not as much as one might have expected. Notice that this is still the expected project profit. We may not know for many months or years after the project has been completed whether our expectations turn out to be valid. But once the project is over, we have considerably more knowledge than we had while it was being performed. We know what the final scope of the product is, we know when it was finished, and we should know how much it has cost: the budget was the amount we expected to invest, the actual cost is how much we have actually invested, and therefore we should now be better able to analyze what we can expect the value of the scope to be.

Experienced project managers will undoubtedly assert that the actual cost will usually be greater than what was budgeted. That certainly is most often the case, which tells us immediately that projects are not being estimated very well: if the budget is an honest target, then surely projects should finish under budget as much as they finish over budget. That we know that this is not currently the case is no doubt a big part of the reason that you are reading this book.

However, the other input to the formula, the project's expected monetary value, has almost certainly changed also. The four factors that will almost certainly have changed the EMV during project performance are

1. Market and other external forces
2. Scope changes
3. Completion date
4. Risk and opportunity changes

The first of those is outside of the project manager's area of responsibility (although if such external forces occur, the sponsor/customer would be wise to inform the project manager ASAP). But all three of the other factors are items for which the project team is responsible. The project manager should manage by

1. Making changes that will maximize the expected project profit (which, we must note, may sometimes mean adapting the plan to those changes in the market and other external forces that the project manager does not control but about which the project manager has been informed)
2. Informing the sponsor/customer and other senior management about internal changes that occur and making recommendations as to how best to adapt to them

The difference in the inputs to the two formulas above should tell us how the project was performed: not only was the actual cost greater or less than the budget, but also has the EMV decreased or increased? And, of some importance, why? Were there external factors that caused the project's product to have less value than planned? Were the original targets for cost and schedule unrealistic or padded? Was the scope of such technical difficulty that, even though technical risks were factored into the original EMV estimate, the reality turned out to be at the far extreme of the risk scale and resulted in extensive rework, scope changes, and delays?

So what we have in those two formulas are snapshots:

1. What did we expect the project profit to be when we initiated the project?
2. What do we expect it to be now that it has been completed (although the actual value of the scope remains to be determined)?

But what is really needed is a metric that can be planned up front that will reflect where we expect the project to be, in terms of expected project profit, at each reporting period along the way. We need a metric that will allow us to track, during execution, our progress toward achieving a project result that meets or exceeds our expected project profit. To do this, such a metric would need to incorporate scope changes (including quality), risk, and opportunity factors, project duration, and cost. And that is what we develop in Chapter 2, where we discuss the third side of the project triangle: TIME.

Examples of expected project profit calculation

EXAMPLE 1: A SIMPLE CONTRACTUAL PROJECT

Our company has been hired by a major multinational corporation to design and build a prototype of a component to be used in a much larger system. The customer needs the prototype urgently and the fixed price plus incentive/penalty contract calls for payment of $1 million if we deliver in 26 weeks, $1.2 million if we deliver in 24 weeks or less, but only $800,000 if we take more than 26 weeks. We have developed a plan with a budget of $750,000 that estimates a 50% probability of delivery in 24 weeks or less, a 40% chance of delivery in 25 or 26 weeks and a 10% chance of delivery after 26 weeks.

What are the expected monetary value and expected project profit on the contract if we deliver precisely on budget?

$$EMV = (50\% * \$1.2M) + (40\% * \$1M) + (10\% * \$0.8M)$$

$$= \$0.6M + \$0.4M + \$0.08M = \$1.08M$$

$$EPP = \$1.08M - \$0.75M = \$0.33M$$

EXAMPLE 2: AN ENABLER PROJECT WITHIN A PROGRAM

The project and contract are the same as in Example 1, but now the prototype development is simply the first project in the customer's program that will require the manufacture of 100 similar components. In addition to the initial contract, there is the potential for additional business from the customer for the follow-on manufacturing contract. This contract will pay $300,000 for each component. Our estimate is that each component will cost $200,000 to manufacture for a profit of $100,000 per component, or $10 million for all 100 components. The prototype contract also stipulates that our company will be guaranteed the manufacturing contract if we deliver the prototype in 24 weeks or less. Although there is no guarantee, our company's management estimates that there is a 50% chance that we will be given the contract if we deliver in 25 or 26 weeks, but only a 10% chance if we are later than that.

Now our prototype project is an enabler of the manufacturing contract. What are the expected monetary value and expected project profit on the contract if we deliver precisely on budget ($750,000) with the $10 million of profit from the manufacturing contract also hanging in the balance? Based on our estimates for delivery of the prototype project, we have a 50% probability of guaranteeing the manufacturing contract (50% of $10 million equals $5 million), a 40% probability of having a 50% chance (= 20%), and a 10% probability of having a 10% chance (= 1%). Suddenly, the EMV of the prototype project is much greater:

$$\text{EMV} = \$1.08\text{M (prototype)} + (50\% * \$10\text{M})$$
$$+ (20\% * \$10) + (1\% * \$10\text{M})$$

$$= \$1.08\text{M} + \$5\text{M} + \$2\text{M} + \$0.1\text{M}$$

$$= \$8.18\text{M}$$

$$\text{EPP} = \$8.18\text{M} - \$0.75\text{M} = \$7.43\text{M}$$

Notice what has happened. It is the direct result of the prototype project being recognized as an enabler project within a program. In this case, the follow-on business value is not guaranteed (few things are in life), but even taking into account the risk of not being given the follow-on contract, the value of the prototype project investment has just gone through the roof! We have to recognize that it is an enabler project or else we are likely to manage it in a way that reduces its value.

When the prototype project seemed likely to generate only $1,080,000 of EMV, we planned it in a way to generate a decent profit by limiting the budget to $750,000. Now that we recognize that there is an additional $7,100,000 of business value riding on the performance of the prototype project, surely we will manage it in a way that does as much as possible to remove any risk of our company not being granted the manufacturing contract. Even if we have to spend an extra $500,000 to guarantee finishing it in 24 weeks, such a decision would be completely justified.

On an enabler project, and even though the revenues from the first project may end up being less than the cost, the loss could simply be an investment to guarantee the additional business value from the rest of the program. We explore this further in the next chapter as we extend the concept to the value/cost of time on an enabler project.

Summary points

1. All projects are investments and therefore a project is only undertaken if the business value it is expected to generate is greater than the expected cost.
2. There is a difference between expected value and actual value. A project's actual value may not be determined until long after the project is finished. But its expected value is based on estimates of the value of the scope if delivered at a certain time.
3. With projects, as with all other investments, the prime operating metric should be the expected monetary value of the project minus the expected cost (budget). Instead, most projects are given arbitrary deadlines and budgets. The project is then performed with the goal of finishing at the deadline and on budget. The result is that opportunities which could increase the business value and expected profit of the project are ignored.

4. The difference between expected value and cost on any investment is called *expected profit*. Therefore every project is undertaken for its expected profit, and that project is managed best which maximizes its expected profit.
5. Not all value from a project comes in the form of revenue or savings. Significant value can often be generated by tangible factors such as patents, reusable technologies, and opportunities for follow-on business, or intangible factors such as customer satisfaction, market visibility, and increased staff expertise.
6. Enabler projects are projects within a program without which follow-on projects would have less value. Such enabler projects must be identified, as their expected value is often much greater than casual analysis would indicate due to the value added to the follow-on projects.
7. Many projects are performed in order to prevent loss of human life. The project scope should be designed to maximize the project's expected value in terms of lives saved.

Endnotes

1. *New York Times*, "As US Agencies Put More Value on a Life, Businesses Fret," by Binyamin Appelbaum, Feb. 16, 2011, http://www.nytimes.com/2011/02/17/business/economy/17regulation.html?_r=1
2. *Time*, "The Value of Human Life: $129,000," by Kathleen Kingsbury, May 20, 2008, http://www.time.com/time/health/article/0,8599,1808049,00.html

chapter two

Of time and timing

> "If time is money, shouldn't we count those benjamins?"

The triple constraint model as introduced in the previous chapter represents an approach to managing a project as an integrated effort. As such, each parameter of the project model interacts with the others. Thus everyone involved in the investment needs to be concerned with all of its key aspects.

Each side of the triangle has special relevance to different functions within the project. Whenever something is relevant to someone, surely they have a vested interest in ensuring that appropriate techniques and metrics are in place.

- SCOPE is of primary relevance to the customer(s) or the sponsor(s).
- COST is of primary relevance to finance and accounting. And, because it is the amount of money that the sponsor must invest to generate the project's value, it clearly is also important to both the sponsor and the project team in terms of working within the target cost constraint.
- TIME, the third side of the triangle, is where the project team has the most autonomy in terms of management.

The project management discipline has developed numerous techniques and metrics for planning and managing project duration. However, most of those who work outside of project management have little knowledge of the requirements for and implications of these techniques. These methods include critical path analysis, resource scheduling and leveling, and even flawed earned-value-based techniques such as the schedule performance index (SPI) and earned schedule tracking (more on this later). Sad to say, those finance departments which frequently wield great power over project performance are seldom adequately trained even in such basics as critical path scheduling. For example, the fact that Activity X, which has a resource cost of $5,000, can actually be far more costly to perform than Activity Y with a resource cost of $200,000, seems absurd to a finance department, but perhaps not to a savvy project manager who knows that Activity X is on the project's critical path!

Simple example of the complexities of managing time

Let us assume that we want to dig a trench two kilometers in length.

- One method would be to hire 10 shovel-bearing laborers who could dig the trench in five 8-hour days. The hourly rate would be $10 per hour. The total cost would be $10 * 10 laborers * 40 hours each = $4,000.
- The other method would be to rent a backhoe and operator for two days, which would cost $2,500 per day. The total cost would be $2,500 * 2 days = $5,000.

Which way would be better?

It depends, doesn't it? The backhoe would cost $1,000 more than the people with shovels. But it would also mean that the trench would be completed 3 days sooner. In order to make a wise decision, we need to know precisely how much those 3 days are worth. If the value of each day is $333.33, then the 3 days will be worth $999.99 and the 10 shovel-bearing workers will be the better investment by one cent. More than $333.33 per day and the time that the backhoe will save would make the extra cost of renting the backhoe worthwhile.

On most projects, computing the time savings of any given resourcing decision is much more complicated, but the principle remains the same: we must know the value/cost of time on the project in order to make intelligent decisions. Time is rarely a fixed quantity: it varies according to the nature of the work, the quality and quantity of resources assigned to the work, and the sense of urgency with which those resources approach the work. Making decisions based on these variables is a large part of the project manager's job.

If I could wave a magic wand and deliver one gift to every project manager and every project team in the world, it would be the gift of information about the cost of time for any given time period of their projects' durations. The projects that are scheduled best, in terms of expenditure of resources and diligence, are those on which there is a confident and well-understood estimate of the cost of time. Prime among these are projects such as nuclear power plant refueling outages ($500,000 to $2 million in the United States, depending on the specific electricity grid) and refineries in oil and gas production, where the cost of every day offline is routinely calculated.

But on the vast majority of projects, from the implementation of new corporate information systems to mergers and acquisitions to construction to pharmaceutical and medical device development to public health vaccination programs to business contract negotiations to emergency responses, the project team whose decisions and urgency can have a vital impact on project duration are almost never provided with a clear estimate of both the value of acceleration and the cost of delay.

Variations in schedule and project duration almost always have a huge impact on the project investment. Furthermore, although there are exceptions (usually in the form of subdeliverables, interim payments, patents, etc.), the value of a project is not realized until it's completed and implemented or, frequently, over the course of many months or years after completion. With exceptions, later completion for a project means that its value would be expected to decrease, as it often causes a later start to value generation and a shorter window before obsolescence. An earlier completion reverses this, providing a larger window of value generation. In addition, factors such as seasonal market windows, first-to-market, and liquidated damages (LDs) can hugely multiply the value impact of project duration. And obviously, in combat, emergency, and healthcare situations, the cost of delay can often be measured in lives lost.

Examples of impact of time on projects

Unfortunately, despite a generalized sense that "time is important," one of the great shortcomings of the traditional project management methodology is the failure to measure and manage the impact of time on the value of the project investment.

Time affects project profit in two ways:

1. *Marching army costs* as mentioned in the previous chapter: the cost of overhead, project management and support, and opportunity costs due to the unavailability of the project's resources for other work as long as the project continues. These costs are associated with resource usage and are added directly to the project costs and cost overruns. Thus the longer the project's duration, the more these factors add to the cost. A general rule of thumb on large DoD programs is that marching army costs are 10% to 15% of the monthly burn rate. So if the finance department says our project is burning through $50,000 per week, every week by which we can shorten it should reduce the final project cost, including overhead, by between $5,000 (10% of $50,000) and $7,500 (15% of $50,000).[1]
2. More significantly, an *acceleration premium* or *delay cost:* increase or reduction in the value of the final product due to change in completion and/or delivery date.

The first of these, marching army costs, are significant enough that every finance department should be keeping both the project manager and senior management updated on them for both the current and future reporting periods (they can change during different phases of large projects).

However, even more important is the impact of delay on the value of the final product. This can be huge. Some examples of these delay costs in different projects are:

- Our previously mentioned remote-controlled Softy the Snowboy needs to be in the retail outlets by no later than the Friday after the Thanksgiving holiday. A single week later might reduce our revenues by as much as 33%. And a five-week delay, until January 2, would reduce revenues to almost nothing; nobody buys Softy the Snowboy after New Year's Day!
- A refueling outage at a nuclear power plant in the United States requires the plant to purchase electricity from the grid at twice the cost per megawatt that it receives when it sells electricity to the grid. This may cost the plant operators anywhere from $500,000 to $2 million per day for every day until the plant powers back up and starts generating electricity again.
- Every day that a new pharmaceutical product or medical device is delayed in reaching the market typically costs many millions of dollars for several reasons:
  - The patent has already been approved, so the clock is running until patent protection expires.
  - A competitor may bring a similar product to market before we can, and order-to-market is a major determinant of eventual market sales: Prozac was able to become an eponymous antidepressant because it was first to market, and Zoloft and Paxil trailed behind.
  - No revenues can be generated until we can start selling the product.

And, of course, people may be suffering and dying for lack of the drug or medical device. Although that collateral damage might not affect the company's bottom line, it is the sort of metric that the development team might want to keep at the back of their minds. If the work is intended to save lives, those on the team can save more lives by finishing the project faster. A project team that knows it is saving lives with each day by which it shortens the schedule is usually a motivated team.

Impact of time on emergency response projects

Government response to an emergency has particularly dramatic delay costs. Tardy reaction by the US Corps of Engineers in response to flooding caused by spring runoff could destroy lives and hundreds of millions of dollars worth of homes, businesses, and farmland. Failure to distribute timely flu vaccines, or a slow response following a major hurricane/earthquake/tsunami/terrorist attack would not only have economic consequences but also vast human costs. Anyone watching the televised scenes of human misery in the immediate aftermath of Hurricane Katrina must have felt great frustration at the lack of timely response. But knowledgeable project managers also felt great anger: whatever else it may have been, the Katrina tragedy was a failure in the management of an emergency response project.

Relief response to such predictable challenges must be planned, optimized, rehearsed, and managed, utilizing standard project management techniques such as work breakdown structures, critical path analysis, resource assignments, and a responsibility assignment matrix (RAM). In the days following Katrina the US Coast Guard was credited with rescuing over 33,500 people from the flooding for which the service was presented with a Presidential Unit Citation.[2] How many more might a more timely response, optimized using project management techniques, have saved? In Chapter 21 of the 2013 CRC Press book *Handbook of Emergency Response* (Badiru and Racz), I described exactly how to use critical path analysis and the new *drag* and *drag cost* metrics (in this case, measured in human lives) to plan template plans for disaster response scenarios. It may be the most important thing I have ever written in my life. I hope it will have an impact on a future emergency response. But the question still must be asked: have we learned our lesson? Has the proper planning been done for the next catastrophe? We can only hope.

Impact of time on enabler projects

In Chapter 1, we discussed the value of an enabler project and how it is kindled by all the other projects that are dependent upon it. In our previous discussion, we emphasized how important it is to recognize that the value of our *Softy the Snowboy* film is much greater than simply its cinema ticket revenues because the movie kindles the merchandise, DVD, and music sales that the film enables.

This kindling effect is often even more dramatic when the schedules of the follow-on projects are dependent on, or are schedule successors to, the enabler project. To continue the example we introduced in Chapter 1, let us suppose that we plan to open our *Softy the Snowboy* movie in all the shopping mall megaplexes on the final Monday in November, with expectations of it playing for six weeks and bringing in revenues as shown in Figure 2.1.

The retail sale of the toys and other merchandise is timed to begin at the same time the movie opens, to include the big Black Friday shopping

	Week 1	Week 2	Week 3	Week 4	Week 5	Week 6
For week	$20M	$20M	$20M	$25M	$35M	$20M
Total revenue	$20M	$40M	$60M	$85M	$120M	$140M

Figure 2.1 Weekly ticket revenues for the *Softy the Snowboy* movie.

day of Week 1 and to continue through the end of Week 8. The revenues for the Softy merchandise will be primarily to families that have taken their children to see the movie and will therefore be affected by total ticket sales and revenues to any given point in time. Moviegoers may not purchase the merchandise immediately after they see the movie, but the marketing department assures us that a certain percentage of those who see the movie will buy merchandise for their children by two weeks after the movie closes. The estimates of merchandise revenue based on movie ticket revenue are as shown in Figure 2.2.

Now let us assume that a delay occurs and the *Softy* movie is unable to open until one week later. The Week 1 ticket revenues totaling $20 million will disappear, meaning the total ticket revenues will fall by $20 million, from $140 million to $120 million. See Figure 2.3. But even worse, every week of merchandise sales will now become a percentage of a lower cumulative revenue number, as shown in the table in Figure 2.4.

The result is that the revenues for the merchandise will be reduced by $88 million, from $401 million to $313 million. If we regard *Softy the Snowboy* not as a movie project but as a program investment, the one-week delay has cost $108 million, reducing total revenues from $541 million to $433 million, $20 million on the movie and $88 million on the merchandise.

Where did the delay occur? It was in the movie project, the enabler not just of the ticket revenues but also of the merchandise revenues, as without the movie none of the other merchandise revenues would be generated. Thus the expected monetary value of the movie project is equal to the value of the entire program, and the delay cost for any delay in the movie project is equal to the delay cost for the entire program!

	Week 1	Week 2	Week 3	Week 4	Week 5	Week 6	Week 7	Week 8
Cumulative Softy ticket revenue	$20M	$40M	$60M	$85M	$120M	$140M	$140M	$140M
Percentage projected for merchandise	25%	50%	75%	100%	100%	75%	10%	5%
Weekly merchandise revenue	$5M	$20M	$45M	$85M	$120M	$105M	$14M	$7M
Total merchandise revenue	$5M	$25M	$70M	$155M	$275M	$380M	$394M	$401M

Figure 2.2 Weekly merchandise revenues for *Softy* as a percentage of cumulative ticket revenues.

	Week 1	Week 2	Week 3	Week 4	Week 5	Week 6
For week	$0	$20M	$20M	$25M	$35M	$20M
Total revenue	$0	$20M	$40M	$65M	$100M	$120M

Figure 2.3 Reduced weekly ticket revenues for the *Softy* movie due to one-week delay in opening.

	Week 1	Week 2	Week 3	Week 4	Week 5	Week 6	Week 7	Week 8
Cumulative Softy ticket revenue	$0	$20M	$40M	$65M	$100M	$120M	$120M	$120M
Percentage projected for merchandise	25%	50%	75%	100%	100%	75%	10%	5%
Weekly merchandise revenue	$0	$10M	$30M	$65M	$100M	$90M	$12M	$6M
Total merchandise revenue	$0	$10M	$40M	$105M	$205M	$295M	$307M	$313M
Original estimate of merchandise revenue	$5M	$25M	$70M	$155M	$275M	$380M	$394M	$401M
Cum. reduction in merchandise revenue	–$5M	–$15M	–$30M	–$50M	–$70M	–$85M	–$87M	–$88M

Figure 2.4 Reduced merchandise revenues due to one-week delay in the *Softy* movie opening.

If a delay had occurred on the merchandise project, sales of the merchandise would have suffered. But as long as the *Softy* movie opened on schedule, its revenue would not have been affected because the merchandise project was not an enabler of the movie project. There are cases where each of two or more projects in a program can kindle one another, and then the impact of each on the other(s), and the impact of any delays on each other, would need to be factored in. Here we are demonstrating just one enabler and one "enabled" project. The cost of the delay on the enabler is the sum of the cost of delay on the entire program, because that is the amount of business value that is being lost.

Impact of time on contractual enabler projects

Not infrequently, contractors take on enabler projects. Usually the customer is acutely aware that the project is an enabler of other projects that are driving the main value of the program. Even if the customer is ignorant of the concept of enabler projects and how to compute their kindled value, he senses that time is very important on the enabler project because other items either cannot go ahead or cannot generate their value until the enabler is completed.

However, even though sometimes senior management and the business area leaders of the contractor organization may be aware of the value-added possibilities of the initial contract, the program manager is almost invariably oblivious of the opportunities, and the project team is completely in the dark. Yet for a contractor, projects for enabler projects can represent lucrative chances to

- Negotiate time-based incentive clauses.
- Position themselves to win the customer's contracts for follow-on work.

Every contractor should be acutely aware of such opportunities. Time is so important to the customer on an enabler contract that it presents a golden opportunity to compete on the basis of faster delivery and generate a win–win situation for both customer and contractor. In the contracting world, customer satisfaction and the reputation for creating customer satisfaction can be very significant value drivers and competitive differentiators. All else (e.g., quality and cost) being equal, shorter projects are an extremely reliable way of ensuring customer satisfaction.

To make good project decisions, it is crucial that the contractor business leaders, the project manager, and the individuals on the project team understand the potentially huge value of finishing the project early. But schedule acceleration does not happen accidentally; it happens because of effort. If the subject matter experts in the work being done on the project's critical path are told the monetized value of acceleration, they will seek opportunities to reduce the drag and drag cost of those activities, often by targeting specific activities for additional resources, with the cost increase more than justified by the added business value. We show precisely how to implement such schedule optimization in the chapter on critical path method.

Ignoring the cost of time

Despite Benjamin Franklin's fame as perhaps the most brilliant of the founding fathers of the United States, on projects his warning that time is money is largely ignored. Why?

People understand that business is about making money, and most understand that in order to make money in business, one must spend money. And the less that can be spent in order to make a dollar, the better the investment and the more dollars remain for other investments. Therefore whole corporate departments, with accountants trained in graduate schools, are employed to keep track of expenditures. They slice them and dice them and estimate overhead 100 different ways from sundown. And that's the way the business world has operated for generations.

By contrast, project management is a very new discipline. Projects have undoubtedly been around since long before the pharaohs decided to construct those buildings that look like the triple constraint model of a project. But the modern techniques we use to manage projects are very new. Most texts date the start of "modern project management" to 1957, with the development of the critical path method (CPM) of scheduling at the DuPont Corporation. As late as the mid-1990s, there were only three universities in North America that offered a master's degree in project management.

Not only is the project management discipline still in the process of developing and discovering new techniques, but the rest of the business world is still in the process of discovering the project management discipline! Outside of those trained in project management, very few corporate employees have been exposed to what a critical path is. That includes human resources departments (who supply and support the resources and skills that the projects need), finance departments (who track and try to control the dollars that projects spend), and senior business leaders who are critical to guiding the focus of project management.

Thus the typical management roles for the three parameters of a project are:

1. Engineers and other technical subject matter experts (SMEs) planning, executing, and managing the scope
2. A department of accountants riding herd on resource usage and cost
3. And, depending on the organization, either
   a. A technical SME trying to snatch an hour a week away from his "real job" to work as a project manager and track schedule, or
   b. A scheduler who is often among the lowest-paid people in the organization and whose time is shared by multiple projects

Yet that schedule, with its impact on both cost and the value of the scope, can have huge implications for the return on the project investment. But despite the lip service that "time is important," the business world still seems to not fully understand either how important project scheduling is nor what needs to be done to turn the current situation from a negative to a positive. The first step to accomplish this is to monetize the value of time on each project.

Time as an externality

Instead, time is left unmeasured, what in economics would be called an *externality*. An externality is any part or by-product of an economic enterprise whose impact is unquantified. Anything that is left unmeasured is usually assumed to be equal to zero. It is well known in economics that if significant aspects of an enterprise, either positive or negative, are left as externalities, they will play no part in decision making and bad decisions may be made as a result. An example of an externality can be seen in the use of antibiotics in the industrial farm animal industry. This has resulted in lower costs for the industry and safer and more abundant meat for consumers. But what is not measured is the healthcare cost to people and society due to the increase in illnesses caused by antibiotic-resistant bacteria.

Therefore on projects, the impact of time on the investment is left unmeasured, while a whole department tracks the dollar cost of the resources that could ameliorate the negative impact of time. No wonder project managers moan about their inability to justify the additional resources that they know they need.

By quantifying the value of time instead of leaving it as an externality, suddenly those tradeoffs between the cost of time and the cost of resources can be made, and the return on the project investment thus enhanced. If the additional work of an extra hobbit for three weeks at an overhead-burdened salary of $1,200 per week will reduce the project duration by one week, the only way to know if the additional investment is worthwhile is to know the value of that one week less of duration. This is

- The opportunity that the project manager and team should be seeking out
- The data they need to determine whether such a tradeoff is worth recommending
- The decision that the project manager should either make (if the tradeoff will not force the project over budget) or recommend to the sponsor/customer/senior management for consideration

Sponsors sometimes seem fearful that knowing the cost of time would somehow empower the project manager and team to make tradeoffs that the sponsor doesn't want. This is completely unjustified, as there is nothing about knowing the value/cost of time that in any way eliminates the need to perform the project within the budgetary constraints set by the sponsor. Knowing the value/cost of time simply assists the project team in looking for those opportunities that sponsors and senior management always seem to emphasize, but never really explain exactly what they are in quantified terms.

If the project manager comes to the sponsor with a suggestion whereby the project schedule could be shortened by a week at the cost of going over budget by $3,600 (i.e., the cost of three hobbit-weeks) because the sponsor said that every week was worth $4,000, I can very easily visualize a smart sponsor saying no. Remember, these numbers are usually just estimates: $3,600 to save $4,000, or $36,000 to save $40,000, is basically a wash.

Conversely, if $3,600 would save a week that is worth $40,000, I think most sponsors would say yes, unless the organization is critically cash-strapped or the sponsor simply doesn't trust the project team to deliver. Either of these reasons may be completely sound, but the key is for the sponsor/customer to have the opportunity to say no. And that only happens if the sponsor/customer provides the data the team needs to assess such opportunities and also encourages the team, in contractual or less formal terms, to engage in such a "treasure hunt."

A few years ago, a woman I know decided to run a personal ad on a dating site on the Internet. She spent a great deal of time phrasing the ad "just right," so that people in whom she might not be interested would not respond. When her ad attracted very few responses, I told her that she had used the wrong approach: she should have cast the net widely and then selectively narrowed the field. Similarly, sponsors should always encourage project team members to propose lots of opportunities to enhance the value of the project investment. Like my friend with the personal ad, there will always be an opportunity to say no. But you don't want a potentially good opportunity to be filtered out too early.

Summary points

1. The failure to monetize the impact of time on the expected value of the project investment, leaving it instead as an externality, is one of the biggest failures of traditional project management.
2. Not only is the value of an enabler project enhanced by the value of all the work it enables, but the value/cost of the time that the enabler project requires is also multiplied by its impact on how soon the enabled work can start creating value.
3. The impact of time can be huge. If it is estimated in monetary units and provided to the project team, then the subject matter experts can seek out opportunities to increase resource costs by a little in order to decrease the business value impact of time by a lot.
4. On certain projects where the end result may save lives, such as in healthcare, pharmaceutical development, or national security, the cost of time can be measured not just in dollars but also in the lives that can't be saved until the project is finished.
5. Providing the project team with the information about the value/cost of time does not empower the team or the project manager to over-spend the project budget in order to compress the schedule. It merely enables the team to regard opportunities they should analyze and perhaps recommend to the sponsor/customer. All decisions regarding changes in scope, duration, and budget remain the province of the sponsor/customer and implementation of any such change requires a formal change in the project baseline plan.
6. The term in economics for a factor that is left unmeasured is an *externality*. The impact of an externality is almost always ignored when cost/benefit analysis is performed. Leaving the value/cost of time as a project externality makes it impossible to justify any expense to reduce its impact.
7. On contractual projects, if time is of significant value to the customer, it is imperative that the customer ensure that the contract include a clause that provides incentive to the contractual team to seek out opportunities for early delivery.

Endnotes

1. Of course, one should take into account whether the cost of those resources to the organization really will end when the project ends. Is freeing up those resources a week earlier simply going to result in an idle week? If so, there

is no value to the organization. Alternatively, if those resources migrate to another project, then the value of the time saved on the first project equals the value of the work performed by the resources on the second project. This sort of analysis is covered well in Steve Jenner's excellent book, *Managing Benefits,* published by APMG-International, 2012.

2. US Government Accountability Office, Report to Congressional Committees, COAST GUARD: Observations on the Preparation, Response, and Recovery Missions Related to Hurricane Katrina, July 2006, p. 1.

chapter three

Tracking projects by investment value

"How are projects measured, tracked, and judged?"

Let us imagine that our company performs lots of very similar projects, what we might call "cookie-cutter" projects, with identical scope: building identical fast food restaurants, for instance. Performance to schedule and cost are the two primary metrics by which every project is tracked. Let us imagine that the planned duration for each of these projects is six months, with a budget of $500,000.

Each of these projects is now finished, with the results as shown in the table in Figure 3.1. Which was a better result?

- Was Project A better, to take two months extra and finish on budget?
- Or was Project B a better result, overspending the budget by $200,000 but finishing at the target duration?
- Faced with being either late or over-budget, would it have been better to do what Project C did and prune some optional scope?
- Or would we have been best off to choose the project D option as soon as we saw that we were headed for big trouble, canceling it after just one month and $100,000 spent?
- Even though Project E went $300,000 over budget, could finishing it a month early have more than made up for the overspending?
- Or could the 10% extra scope we included in Project F have made up for both the extra month and the overspending?

These projects are finished, and although we know how each did in the performance metrics that project management traditionally uses, we still don't know which results were better and which worse. Yet we achieve such results not by accident but by management decisions! How can we possibly make the decisions that would lead to any of these results if we don't know which would be better?

Expected value versus actual value

One common objection that is often raised about tracking a project as an investment is that even when we finish a project, we often do not know what its value will be. "The value is generated long after the project

Proj. Name	Duration	Cost	Scope
Proj. A	8 months	$500,000	100%
Proj. B	6 months	$700,000	100%
Proj. C	6 months	$500,000	80%
Proj. D	1 month	$100,000	0%
Proj. E	5 months	$800,000	100%
Proj. F	7 months	$800,000	110%

Figure 3.1 Results of six completed projects of similar scope.

is finished," the objectors say. "Frequently, the value is determined by factors that the project manager and team have absolutely zero control over, such as the sales-and-marketing campaign, or poor training and support that causes the new information system to never generate the value it should. The project team may have done an amazing job; it's not their fault that after they were done, the investment crashed and burned!"

Let me stipulate that this is absolutely true. It is also irrelevant. The metric that should be used for tracking a project is not the project's value, but the project's expected value. And, as with all investments, events beyond our control can cause projects to crash and burn. But we should always know what the expected monetary value of our investment is: what revenues, savings, or other values we are expecting to generate and which factors are within the project's control and may change its value from what we expected.

In the case of a project investment, those factors are the ones that are the responsibility of the project manager and the team: (1) SCOPE, (2) COST, (3) TIME, and (4) RISK. These are the items that must be planned and approved, and optimized on the basis of generating the most expected profit for the project investment.

At the start of the project, we should know:

1. The value we expect to get from the planned scope if the project is completed on the target finish date.
2. How much value earlier completion would add, and how much value later completion would subtract, for each significant calendar period. (Note that this may not be a straight line function: the first week later or earlier might have a bigger impact than additional ones, or vice versa.)
3. The *planned cost estimate-to-complete* (ETC) at any date during the project. At the start of the project, the planned cost ETC is identical to the budget. But as the scheduled work is performed and paid

for, the planned cost ETC should decline by the amount that was budgeted for the work already performed. Figure 3.2 shows the planned cost ETC function, as the complement of the planned cost accruals function (which, as we show in the chapters about earned value tracking, actually has three other names: it is also called the *planned value* (PV) or the *budgeted cost for work scheduled* (BCWS) or the *earned value baseline*). As the cost accrual curve is planned to rise, the cost estimate-to-complete function is planned to decrease by exactly the same amount.

4. Specific identified risk factors that might cause any of the above to be different from what was planned. All three of the above parameters (i.e., expected value of the scope if completed on a certain date, acceleration premium or delay cost, and planned cost ETC) should be estimated, with appropriate modifications for risk. And any risk factors that can be scheduled to be retired at a certain point during project implementation (for instance, the risk of incurring the cost and delay of rework being retired after successful testing) should be tracked, and the impact of any such risk retirement incorporated into the project investment metrics and the expected monetary value of the project.

As shown in Figure 3.3, all three sides of the triple constraint model are now quantified in dollars. This provides the basis for a formula that can be used to measure the expected project profit (EPP) of

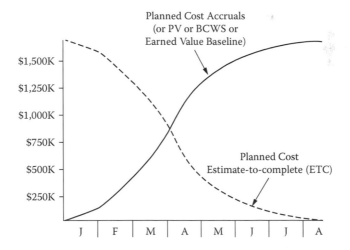

Figure 3.2 Planned cost ETC function as the complement of the planned cost accrual function.

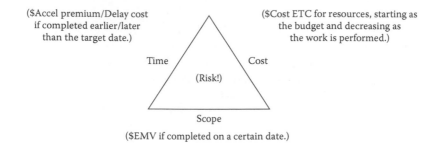

(\$Accel premium/Delay cost
if completed earlier/later
than the target date.)

(\$Cost ETC for resources, starting as
the budget and decreasing as
the work is performed.)

Time

Cost

(Risk!)

Scope

(\$EMV if completed on a certain date.)

Figure 3.3 The monetized version of the triple constraint model.

the investment after the project is over, when we know its results in terms of duration and cost:

$$\$EPP = \$EMV \pm \$acceleration/delay - \$Cost\ ETC$$

where:
1. The EMV is generated by the SCOPE (modified by the RISK of uncertainty) if completed on a given date.
2. It is modified by the added or subtracted value/cost of TIME if completed earlier or later.
3. The COST of the project's resources (the investment!) is subtracted.

Let's look again at the six identical projects (A–F) in Figure 3.1. For all of these projects, let us assume:

- An EMV of \$1 million if completed in six months
- An acceleration premium of \$400,000 for each month earlier
- A delay cost of \$450,000 for each month later
- A reduction in EMV of 25% for the reduction of scope in Project C
- An increase in EMV of 35% for the added scope in Project F

Now we can get a better picture of which projects were better investments and which were worse. See Figure 3.4.

- Project E is the best result by far, generating twice as much expected project profit as Project B due to its taking advantage of the \$400,000 acceleration premium (for which it pays \$300,000 in resource costs).
- Although Project D's cancellation results in a \$100,000 loss, whereas Project C prunes scope to a value of \$250,000, both of these results are better than for Project A, which stuck rigidly to both budget and scope parameters and therefore had delay costs of \$900,000 for being two months late, resulting in an EPP of minus \$400,000.
- Project F's team may have assumed that the \$350,000 of value of the additional scope would only cost \$300,000 in resources. However, if

Proj. Name	Duration	Cost	Scope	EMV	Scope±	Time±	EPP
Proj. A	8 months	$500,000	100%	$1,000,000	$0	−$900,000	−$400,000
Proj. B	6 months	$700,000	100%	$1,000,000	$0	$0	$300,000
Proj. C	6 months	$500,000	80%	$1,000,000	−$250,000	$0	$250,000
Proj. D	1 month	$100,000	0%	$1,000,000	−$1,000,000	$0	−$100,000
Proj. E	5 months	$800,000	100%	$1,000,000	$0	$400,000	$600,000
Proj. F	7 months	$800,000	110%	$1,000,000	$350,000	−$450,000	$100,000

Figure 3.4 Table of six projects including scope changes and the value/cost of time.

that scope addition shifted the critical path and caused the one-month schedule delay, then the true cost (TC) of the additional scope was actually $750,000: $300,000 of resource cost plus $450,000 of project delay cost. Yet without recognizing the cost of the added project time, the $300,000 of additional resources might have seemed worthwhile. This example demonstrates the importance of always analyzing the impact of scope changes on both cost and schedule, and of monetizing the schedule impact through previously determined estimates of acceleration premium or delay cost.

To re-emphasize: as with any investment, the expected project profit may turn out to be very different in fact from what was estimated. It is entirely possible (although we hope unlikely) that a completely unforeseeable risk (what Nassim Nicholas Taleb referred to as a "black swan"[1]) may affect the fast food restaurant that we built in Project E and perhaps obliterate both the structure and its expected value. But that does not excuse a failure to try to take into account those investment factors we can control, that is, the project management parameters that are likely to lead to the most profitable investment. And without clear estimates of the value/cost of time as well as both the value-added and the true cost of scope additions and subtractions, we are back to managing our project in an unlit room on a moonless night.

A final note on expected project profit versus actual project profit: the delta between these two numbers on every project is an extremely important metric for any organization to track. A senior manager who repeatedly sponsors projects on the basis of inflated expectations of value and profit can fritter away large amounts of corporate revenue. There are two different ways that a sponsor might do this:

1. Overestimate the value/revenues/savings of the product, service, or result that the project produces
2. Set scope/time/cost parameters that the project is extremely unlikely to be able to achieve

It should be critically important to any organization to gather and track metrics to determine just how well each senior manager is performing in his or her role as project sponsor.

1. Does one vice president periodically fund new IT systems, promising that the reductions in headcount and processing time thus generated will result in $2 million per year in savings, only to have the new system generate less than 10% of that value?
2. Do a particular sponsor's project parameters frequently use "lowball" estimates, saying, for example, that a project should produce expected project profit of $5 million over three years provided it is completed in six months for a $2 million budget, only to have the project invariably take much longer and cost much more?

Both of the above situations can do serious harm to an organization if allowed to recur frequently. In many ways, the second scenario may be worse than the first, as it can have an extremely negative impact on project team morale.

Under any circumstances, project sponsors should be held as responsible for their estimates as is a lowly engineer, probably even more so, as there is usually more money at stake. Please do not misunderstand my intention: estimates are but estimates, and some of them will be wrong. Estimating skills can be improved, provided that negative estimating tendencies are identified and measured. But this requires documentation of both estimates and actuals, and measuring the difference between them. Sponsors whose estimates are consistently inaccurate and who fail to improve over time can be a huge drain on the organization.

DIPP: A formula to analyze termination of a project investment

In 1990, at a time when I had been teaching project management theory for corporations for about two years, I found myself reading many project management articles on the subject of how to know when an organization should continue funding a troubled project and when it should terminate it. One Sunday night I played in a backgammon tournament. As I was driving to work the next morning, the thought occurred to me that the continue-or-terminate decision point on a project is not that dissimilar from a backgammon decision: when an opponent doubles the stakes of the game and you have to decide whether you're better off conceding the original stake or playing for double the amount.

- In backgammon, your opponent will double the stake only if he is winning, so by continuing to play for the doubled amount, you are accepting these terms despite a losing position.
- However, if you refuse to play for the doubled stake, you thereby accept with 100% certainty the loss of the original stake.

The correct decision rests not on which way will you win money or points because here you are either certain to lose the original stake or are probably going to lose double that amount. The real issue is, which way will you lose less money?

Although there are complicating issues that can make the backgammon decision much more problematic, it turns out that the key question is: do you have more than 25% equity in the position or less than that? To demonstrate why, let us assume that you play four games, each for one dollar, in which you reach a position where you have exactly 25% equity, and in each case your opponent doubles the stake at that point. Should you "drop" her double or "take" it?

- If you drop her double and concede the original stake each time, you will lose a total of $4.
- If you take her double, you will lose 75% of the four games, each of which is now worth two dollars, for a loss of $6. But you should also win 25% of the four games, or one game worth $2. Losing $6 and winning $2 also results in a $4 loss.

With exactly 25% equity, you will lose an identical amount whether you drop every double or take every double. With more than 25% equity, you will do better by taking every double; with less than 25% equity, you will do better by dropping every double. Notice that you lose money either way, but when you're going to lose money, your task is to find the way to lose the least amount of money.

As I was driving to work that Monday morning in 1990, it dawned on me that none of the articles I had read that discussed the project termination decision touched even tangentially on this value/cost aspect. I therefore wrote an article proposing a formula (which I called the DIPP, for Devaux's Index of Project Performance) for guidance in making such a decision. But the article's main concepts, that you should factor out a project's sunk costs when considering a project for termination and that you should keep funding it provided that its expected value is greater than its cost estimate-to-complete (Cost ETC), seemed so simplistic to me that I hesitated to send it to any publisher. Finally, my boss insisted that I submit it to the *Project Management Journal*, the refereed magazine of the Project Management Institute (PMI). I fully anticipated a brief response to the effect that my ideas covered old and well-worn ground. Instead, one of the referees said it was excellent stuff and should definitely be published. The other referee said that it should be made the featured article in an upcoming issue.

And that is when it dawned on me that the theoretical basis for our project management discipline was still in its infancy; thus a concept that would have seemed simplistic even to novice backgammon and poker players was the basis for an article in probably the most prestigious of project management publications. "When the DIPP Dips: A P&L Index for Project Decisions" was published in the September–October 1992 issue of *Project Management Journal* and then reprinted as the featured article in 1999 in Pinto and Trailer's compilation of theoretical articles titled *Essentials of Project Control*,[2] published by the Project Management Institute.

The DIPP formula was stated as follows:

$$DIPP = (TPCM - OC - CW) \div (Cost\ ETC - PTC)$$

where TPCM is the total project contribution margin (the equivalent of EMV), OC is opportunity costs, CW is cannibalization worth (or salvage value), PTC is project termination costs, and all elements are adjusted for risk and discounted by a common time cost of money factor. If the numerator is greater than the denominator, the DIPP will be greater than 1.0, suggesting that there will be greater value in completing the project than in canceling it. If the DIPP is less than 1.0, that suggests the project should be canceled unless the plan for the rest of the project can be changed so as to generate a DIPP for the remaining work that is above 1.0.

Obviously, as a project nears completion and its cost ETC decreases toward zero, the project's expected monetary value can become less and less and its completion will still be financially warranted. If a project with a budget of $1 million starts with an EMV (or TPCM) of $2 million, even though market factors may cause its EMV to severely decline, as long as the cost ETC is substantially less, continued funding is still the wise policy. When the cost ETC is only $1,000, an EMV of just $1,100 may be adequate to justify completion, representing a return of 10% on the remaining investment.

In my opinion, no project should ever be canceled without first performing the full DIPP analysis, and any troubled project (or even one where project performance is proceeding smoothly but the potential market for the future product is in decline) should be subjected to such analysis. In the corporate world, projects are often funded when they should be canceled and canceled when they should be funded to completion. Not infrequently, both mistakes are made on exactly the same project.

1. After three months, a project's EMV drops from $1 million to $250,000 and its cost ETC is $300,000. If there are no other data, it should be terminated as its expected monetary value is less than what it will cost to complete it.

2. Yet its funding continues for another three months during which its
   EMV drops to $200,000, but its cost ETC has declined to just $150,000.
   After the additional three months, continued funding is now $50,000
   better than termination.

The above situation is not unlike so many novice poker players who
decide to "pay to see one more card" when they'd be better off folding,
and then fold when the minimal price remaining to call the hand is more
than justified by the size of the pot. But even journeyman poker players
in their neighbor's Saturday night game quickly learn to avoid that trap.
How is it that so many corporate executives don't?

In February 2003, an article appeared in the *Harvard Business Review*
titled, "Why Bad Projects Are So Hard to Kill" [3] by Isabelle Royer. Royer
focused on large projects at two different corporations, Essilor and
Lafarge, which she suggests are exemplars of projects that should have
been canceled long before completion. But nowhere in the entirety of the
article does Royer ever provide the reader with the two crucial data items
on which such a decision would need to rest: the expected monetary value
of the project and its cost estimate-to-complete. Without those two items,
there is no evidence that either project ever reached a DIPP of less than
1.0 and should therefore have been terminated, or that Essilor or Lafarge
ever made either good or bad decisions. This article, in America's most
prestigious business journal, underscores the gross lack of understanding
of project tracking and decision-making.

Simple DIPP: Setting the baseline for expected project profitability

In the years following the original publication of "When the DIPP Dips,"
it slowly dawned on me that, although the original DIPP formula was a
valuable tool for analyzing whether to continue funding a project or to kill
it, a variant of the original formula could also serve as a project management tool for tracking and optimizing project performance on the basis of
investment value. In my 1999 book *Total Project Control: A Manager's Guide
to Project Planning, Measuring, and Tracking*, I proposed what I have since
called the *simple DIPP* or the *tracking DIPP*. The tracking DIPP is designed
to measure and integrate the three sides of the triple constraint model into
a single index, the essence of which is the business value of the project
investment. The formula is:

$$DIPP = (\$EMV \pm \$acceleration/delay) \div \$Cost\ ETC$$

In the baseline version of this formula (which should be saved at
the start of the project in order to track actuals against it), both values

in the numerator are assumed to remain fairly constant throughout the project, whereas the cost ETC is planned to decrease as work is completed.

1. The project's EMV should be predicted to remain unchanged throughout the project except for the case of a date on which a specific project risk factor is scheduled to be retired. For example, we may estimate that there is a 20% chance that our prototype will fail the test scheduled for August 1, resulting in a decrease in project profit of $50,000. The project EMV should therefore be discounted by $10,000 (20% of $50,000) until August 1 and then increase by $40,000 if that risk factor is scheduled to be confirmed or nullified thereafter. However, it is important to recognize that the EMV may change due to factors beyond the project team's control, such as a global recession or a competitor entering the market with a similar product. Thus the project sponsor should ensure that either the marketing department or some other entity is providing updated data on the project EMV throughout project execution.

2. At the start of the project, there should be no anticipated acceleration premium or delay cost, as the project should be scheduled for completion on a specific target date.

3. However, the third side of the triangle, the planned cost ETC, which starts as the project budget, can be predicted to decrease at a scheduled rate as project work is performed. As we discussed earlier in this chapter and showed in Figure 3.2, the cost ETC can be planned up front and scheduled to decrease as the complement of the planned cost accrual function.

If the EMV is expected to remain constant during project execution and we are not anticipating either schedule acceleration or delay, it means that the numerator of the tracking DIPP formula will remain constant. But the cost estimate-to-complete is planned to decrease during execution, at a rate that we can forecast.

If the numerator remains fixed while the denominator decreases at a predictable rate, we can create a tracking DIPP baseline, displaying where the DIPP is planned to be at every reporting period, as shown in Figure 3.5. The fact that the DIPP is scheduled to rise steadily, from 2.65 at the project's start to 36.73 at Week 26, is neither a mistake, nor an aberration, nor a flaw inherent in the instrument: it is reality. As time passes on a project and the cost ETC decreases, future money to be spent to complete that project really does become a better and better investment. Imagine that during Week 18 we receive word from the marketing department that, previous estimates notwithstanding, the EMV for this project has been halved. The DIPP has fallen from 6.03 to 3.00. Completing this project is still almost certainly the right investment: how many other ways are there to get better than a 300% return on the remaining funds we need to invest?

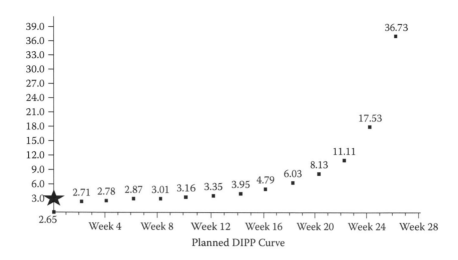

Figure 3.5 The baseline function of the tracking DIPP as the planned cost ETC decreases.

As projects get closer and closer to completion, we need to invest less and less in order to pocket their value. This should suggest that projects which are nearing completion should usually have top priority for needed resources. However, in my experience, precisely the opposite is the case, especially when the project is being performed by a contractor. All too often, the emphasis switches to "migrating" the resources to new projects. This is especially true in organizations where one of the metrics being used to judge the performance of functional managers is utilization rate, that is, the percentage of time that resources are performing billable work. This metric, which contracting companies try to keep at as high a percentage as possible in the belief that it lowers overhead rates on their contracts, is one of the more destructive examples of the saying: "Be careful what you measure; you may get it!"

The following are some of the destructive ways by which functional managers sometimes keep their utilization rates high:

1. Keep staffing levels low, thus ensuring that projects are always desperate for that manager's resources.
2. Make sure that all the resources are multitasked and multiprojected, so that when one task or project finishes there is always another on which to work. (Of course, the deleterious effects of multitasking on productivity, when a resource is constantly being forced to "pull his head" out of one task and plunge it into another, is well known. Not that knowledge of this fact seems to inhibit companies from doing it.)

3. Require that project managers, if they want a particular resource during the last two weeks of July, accept that resource onto their project July 1st because: "Hey, Henry is available at the start of the month. If you don't take him then, I'm going to have to find another project for him and you may not get him until September!" The cost of the project goes up, but who cares? Just as long as that utilization rate on which the functional manager is measured does, too, right?

4. As mentioned above, move resources from projects that are nearing the finish line to new ones. So what if it means that the first (and now very valuable from an investment perspective) project gets delayed by a month? That just gives the functional manager an opportunity to find a third project to which to assign the resource.

It is not my intention to denigrate either the ethics or the conscientiousness of functional managers. Their organization has a whole accounting department tracking overhead costs but failing to measure much more important project metrics: the value/cost of time and the factors that are contributing to that cost. The functional managers are being told that a high utilization rate is of great importance for reducing overhead costs, and, of course, it is, all else being equal, which it rarely is! They are doing their best to obey the directions they've been given.

The fact that this requirement introduces substantial moral hazard is bad enough. But it is almost always hugely detrimental to project investments where all else is not equal; time, work, and resources on the critical path delay project completion and thus are hugely expensive. And the situation is exacerbated by contracts that fail to align the interests of contractor and customer.

If the interests of sponsor/customer, contractor, and project team are all aligned, with the contract reflecting customer benefit by incorporating incentives for the contractor based on better DIPP performance (and perhaps penalties for worse DIPP performance), then:

1. Achieving or surpassing the DIPP will provide the sponsor/customer with the greatest value for the least cost (i.e., most expected project profit).

2. The contractor will also have clear reason to try to maximize the DIPP.

3. The project team, instead of trudging along to deadline and budget goals that may be either impossible to achieve or much too easy, but whose inclusion in the contract often drives counterproductive behaviors (such as excessive risk-taking, quality reduction, or letting the schedule slip until the deadline), will try to maximize its performance at every progress-reporting point in the terms on which it is being measured: the tracking DIPP.

Of course it is helpful if the project team has reason to believe that some of the contractual incentives will be distributed to those team members who enabled their achievement. But in my experience, teams will strive toward outstanding achievement in whatever metric they know they are being measured, not unlike the way that the on-base percentage (OBP) of major league baseball players shot upwards shortly after that was recognized as an important metric in the mid-1980s.

DIPP Progress Index (DPI): Tracking project value against baseline DIPP

In 2002, I was teaching a class of engineers who worked in the space division of a major aerospace contractor in Manassas, Virginia. As I described the DIPP and how it was designed to work, one of the engineers, named Hugh Miller, interrupted me.

"Why don't you normalize the DIPP?"

I was too dense to understand what he meant.

"You know, divide the actual DIPP by the planned DIPP to get an index where 1.0 is on plan, anything lower is worse than planned, and anything higher is better. Just like you would with the cost performance index in earned value cost tracking."

I felt like a moron. I had been working with the concept of the DIPP for more than a decade, and yet I'd never thought of this. I thanked Hugh Miller for his idea, promised that I would always mention his name whenever I talked about this important enhancement to my original idea, and ultimately I started calling his index the *DIPP Progress Index* (DPI).

$$DPI = Actual\ DIPP \div Planned\ DIPP$$

In fact, to refer again to Figure 3.5, if the team knows that at the end of Week 18 the DIPP is supposed to be 6.03, and understands that the quantified value/cost of time is an important element in the DIPP, it will seek opportunities, usually through time/cost tradeoffs, to have a DPI not of .91 but of greater than 1.0. I have seen teams where exempt employees working on the critical path decided to come to work on weekends simply so that their successor activities could start at the beginning of the new week and the DIPP and DPI would be raised through the acceleration premium. And isn't that horrible? That project teams should be inspired to act in ways that create more value and benefit for their employer and for the customer?

Many years ago a program manager from Pennsylvania named Joe Sopko was working for a large telecom firm. He was put in charge of a very troubled product development project, one that had already long since passed its intended delivery date. Right about this time, Joe attended a course that I was teaching in Baltimore for the Project Management Institute's Global Seminars Series. In that course, I described how to use the DIPP to incorporate the cost of time in tracking project value.

Joe returned to his company and immediately went to visit the marketing folks. He persuaded them to do some analysis to quantify the impact of the passage of time on the expected revenues of the new product. He then took that information back to his engineering team. Their reaction was amazement; although they knew the time was supposedly important, no one had ever explained to them that it was actually many hundreds of thousands of dollars per week. Suddenly, progress meetings became both much more urgent and much more collaborative among the team members:

"You say it's going to take you three extra weeks to get that done? That's gonna cost about $2.5 million! Surely there's something we can do. How can we help?"

Notice, this project was already way behind schedule; there would never be any time- or DIPP-based incentives for the team. But just the desire of human beings to do the best job they can in whatever terms their performance is being measured was driving these engineers to make every attempt to "finish strong."

Joe Sopko reports that the team ultimately pulled in the planned end date considerably from what was projected when he started sharing the cost of time, and the final DIPP was considerably higher than when he had taken over the project.

Case for this new approach to planning projects

Human beings tend to be resistant to changes in the approaches and techniques that they have used in the past. Among the nonengineering industries that are new to project management techniques (e.g., pharmaceuticals, information systems, software, and publishing), there has been resistance to adopting even intuitively valuable techniques such as work breakdown structures, critical path scheduling, and earned value tracking.

Project managers are often insistent that their job consists of no more than delivering the mandated product on a specific date and for a specific cost. I believe that this minimalist approach leads not only to inefficiencies and waste, but also to a long list of the problems that the discipline continues to experience. It is no secret that senior managers in every industry are frustrated with project performance that consistently delivers poor projects with what seems excessive cost and schedule overruns.

There is an urgent need for change. I believe that it must start with our very definition of what a project is, a definition that is often understood by senior management and project sponsors but all too rarely by project teams who find themselves struggling to meet the requirements of technically complex work. But everyone in the business world understands the concept of investment. Senior management simply has to start defining projects in those terms, and standardizing processes that focus on business value and how it is affected by the key aspects of a project: scope, time, resources, and risk. Incorporating each of these in its interaction with the others into the organization's standardized value metrics for the project investment has the power to transform the way that projects are perceived, planned, and performed.

The steps of the transformative process are as follows:

- If a project is an investment, then its prime operating metric must be the value it's intended to generate for the sponsors/customers who provide the funding. Value generation above cost, or return on investment (ROI), or profit, is how we judge every other investment. Why should project investments be the one and only exception?
- Once an organization starts to define its projects in this way, the approach to planning and measuring projects will of necessity change to focusing on the value of the project investment. Metrics that reflect the impact of detailed scope, time, and risk will start to drive project team performance and decisions will be made with reference to such metrics.
- The tendency, discussed in Chapter 1, to categorize the outcome of projects in a binary manner, as success or failure, is a meaningless exercise and will become extinct. The success of project investments will be seen as relative and based on project decisions: better or worse than expected, better or worse than an alternative investment, better or worse than it might have been.
- The project data parameters that interact to produce the expected project profit (EPP) are the three sides of the triple constraint model (scope, time, and cost) plus risk, which can affect any of the previous three. Risk identification and management is hugely important on projects, but is often performed in a cursory manner. Focusing on the impact of risks on the prime metrics of business value and project profit provides risk management with a clarity and an urgency that is currently seldom present.
- The investment value to the sponsor/customer and the investment value to a contracting organization performing a project may be different. It should be of crucial importance to the sponsor/customer that the terms of the contract ensure that sponsor/customer benefit and contractor benefit are aligned. Any lack of congruence risks a divergence of goals and can generate moral hazard for the contractor. Once again, recognition of customer business value will drive both parties to contractual clauses which, through incentives, relate the contractor's business value to maximizing the customer's business value. And that is a win–win for all parties.

- The fact that, on the vast majority of projects currently, the value/cost of scope and time are not quantified in monetary units makes them externalities, valued at zero. Both additional scope and compressed time are usually enabled by additional resources whose cost is always measured. Monetizing the value of time will allow justification of such additional resources as will enhance the project investment.
- The fact that the project's investment value can be measured and tracked through expected project profit (EPP), the DIPP, and the DPI allow for continuous improvement during project performance. Opportunities to improve business value that were not visible earlier in the project will surface in response to the project team looking for them, and the project value metrics that are now at the core of progress meetings will quantify their value and justify their implementation.

In the next two chapters, we explore powerful new metrics that allow the project team to identify opportunities to enhance the project's business value: critical path drag, drag cost, and true cost of work.

Summary points

1. Project metrics that do not reflect the integrated totality of the investment value across all parameters give a distorted picture and can lead to bad decisions.
2. As with any investment, the actual value of a project may not be revealed for a long time after the investment is made. However, a project should always be managed in such a way as to maximize the expected profit of the investment.
3. On troubled projects, the effort often becomes not how to make the most money but how to lose the least.
4. As a project nears completion, the fact that its cost estimate-to-complete has decreased means that it often represents a hugely valuable return for the small investment necessary to complete it.
5. Any project that is being considered for termination should be subjected to careful analysis using the original complex DIPP formula.
6. Establishing procedures to track projects using the tracking DIPP and DPI formulas allows the organization to measure project performance based on investment metrics and allows the project team to make decisions that will optimize the project's investment value.
7. On projects performed on a contractual basis, it is in the best interests of both contractors and customers that the contract terms be aligned so as to reflect optimum business value.

Endnotes

1. Nassim Nicholas Taleb, *The Black Swan: The Impact of the Highly Improbable.* New York: Random House, 2007.
2. Stephen A. Devaux, When the DIPP dips. In Pinto and Trailer (Eds.), *Essentials of Project Control.* Newtown Square, PA: Project Management Institute, 1999, 129–141.
3. Isabelle Royer, Why bad projects are so hard to kill. *Harvard Business Review,* February 2003. http://hbr.org/2003/02/why-bad-projects-are-so-hard-to-kill/

chapter four

Managing project time

"What's so critical about the critical path?"

Several years ago, I was teaching a seminar to senior managers in a very project-driven engineering company where most of the attendees had come up through the ranks, serving previously as project engineers and managers. I had begun by discussing how every project is an investment, and how to manage the metrics of such an investment. Then I moved on to the subject of this chapter, using critical path analysis to manage the investment impacts of the project schedule. As I do shortly, I focused on the missing metric in critical path analysis: critical path drag. I defined and described it, why it was so important, and explained how to compute it. Several of these executives were astonished to discover a simple yet crucial new concept in the method of scheduling they'd been doing for years and thought they knew backwards and forwards. Finally, one of the senior managers interrupted me: "Y'know, Steve," he drawled, "you have what I'd call an amazing grasp of the glaringly obvious!"

I hope that much of what was covered in the previous chapters now seems to the reader to be "glaringly obvious," or at least, in retrospect, straightforward and not particularly controversial, even if not standard practice. But what I describe in this chapter is amazingly obvious. The fact that it is not standard practice—indeed, the fact that it has not been standard practice for the 50-plus decades that critical path analysis has been around—is mind-boggling.

For the past 15 years, critical path drag is the topic about which I am most often asked to write articles or make presentations. This is because its value is the most intuitively obvious of all the techniques and metrics that are discussed in this book. When computing critical path drag and its corollary metrics becomes standard scheduling practice, and is supported by all project management software packages, crucial techniques that are almost never done today will suddenly become commonplace. Project teams will finally start to

1. Compress project durations and thus make projects more profitable.
2. Identify ways to recover slipping project schedules quickly and efficiently.
3. Identify both the work activities and the resource constraints that are delaying project completion.

4. Measure how much such work activities and resource constraints are costing by reducing project profit.
5. Justify the additional resources that would reduce such delays.
6. Establish organizational staffing levels that are more efficient in terms of resourcing critical paths and thereby increasing project profits.

I suspect that after you have read this chapter, you will be as dumbfounded as I am as to why critical path drag was not established as a critical path metric decades ago. I have no idea.

Critical path is—uh—critical!

The fundamentals of critical path analysis have been around since the late 1950s. Because the technique is simply a common-sense approach to scheduling work, it's undoubtedly the case that many people in previous centuries, likely going back before the pharaohs and their pyramids, used an approach for sequencing project work that we would recognize as the critical path method (CPM). However, the array of specific metrics computed by critical path analysis (such as total float and free float) probably was not defined until the 1950s.

An argument can be made that no single technique is more important within the project management discipline. Yet in the world of business and government projects, critical path scheduling is often undervalued, or even completely unutilized. And it certainly is poorly understood. There undoubtedly are many reasons for this. But I suspect that a significant cause is the omission, from both project management theory and from scheduling software, of the two vital metrics, drag and drag cost, that we discuss in this chapter. I believe that the understanding and use of these will greatly increase the appreciation of critical path analysis among the professional project management community.

Why is it important that the project management community use critical path analysis? Because, most simply, the critical path determines the length of every project. Just as a chain is as weak as its weakest link, a project is as long as its longest path. Considering this, the fact that some project managers elect not to use critical path scheduling can only be attributed to a lack of understanding. I have heard project managers say, "Well, we don't know exactly what's going to happen, so there's really no point in planning a critical path schedule." Yet every project has a critical path.

Perhaps what causes some to ignore critical path planning is the fact that the initial critical path schedule will almost never be achieved in actuality. Technical problems will arise, expected resources will be absent, there will be a gap in "work hand-offs," and the critical path itself will quite likely migrate to a completely different set of activities. But none of this lessens the benefit of critical path analysis; indeed, the more variances from plan that there are, the more valuable critical path analysis is for measuring impacts and making decisions.

It is important to recognize that the critical path may very well turn out to be different from those linked activities that one expected to comprise the longest path; that path was simply the planned critical path. That said, the project will still be exactly as long as its longest path! The longest path that ultimately determines the length of the project is what in the construction industry is known as the as built critical path (ABCP). This path, with its work delays, technical difficulties, scope changes, and resource insufficiencies, is what ultimately determines the length of the project. And whether it is what was planned or not, it is crucial that the project manager recognize the overwhelming importance of this path, and manage it. During the project postmortem (a vital process at the end of every project that is all too often omitted), this path and the changes from plan that may have triggered it should be a vital artifact. After all, whether the project schedule was ultimately delayed or accelerated, the critical path has almost certainly had an impact on the expected monetary value of the project investment.

It has become fairly common to hear someone refer to a project's "technical critical path." In so doing, they are demonstrating how little they know about project management; the critical path is a function of the project schedule and is oblivious of the technical nature of any of the work. One may have technically challenging activities, or risky activities, or volatile activities, or mandatory activities, but none of that has anything to do with the critical path.

The longest path is called the critical path because:

1. It determines the length of the project.
2. Any delays on the longest path make the project longer; delays that are not on the longest path do not make the project longer unless the delays are so great as to make that path into the longest path.
3. If you want to shorten the project duration, you can only do so by shortening the critical path. One can shorten a path down to a nanosecond, but if it isn't the longest path, the project duration will remain unchanged.

Reason #3 is the one that most people overlook, because the truth is that project managers and schedulers in most industries make almost no effort to compress the schedule provided that the finish date is within that often-arbitrary deadline. Yet from the previous chapter we know that the passage of time almost always has a negative impact on the value of the project investment. Failure to use CPM to compress schedules has a cost.

Much project management literature points to the "birth of modern project management" as having taken place in 1957, when James Kelley, Jr. of the Rand Corporation and Morgan Walker of Dupont worked out the basic techniques and metrics of CPM. Thus the project management discipline is closely tied to the recognition of the importance of critical path analysis. It is therefore quite depressing to look at the histogram in Figure 4.1 generated by the Google Books Ngram Viewer for the number of appearances in books of the terms "project management" and "critical path" between 1955 and 2008. From the diagram, one can indeed see that the two terms started to appear pretty much simultaneously in the late 1950s. The term "critical path" appeared in books far more often than the term "project management" for about 20 years, until the late 1970s. At that point, the usage of "project management" took off, stimulated in large part by the arrival on the scene in the late 1970s of that commercial product known as project management software. The Google Books Ngram Viewer depiction of the volume of usage of that term from 1975 to 2008 is shown in Figure 4.2.[1]

With greater computing capacity and speed, the functionality of project management software has expanded over the years. What do such products do? Coded work breakdown structures, resource leveling algorithms, cost and earned value plans and reports, and Monte Carlo systems for analyzing risk, all have been added to even basic project management software products. But the one core functionality that almost all these packages have is what is called a CPM algorithm, which computes the duration of the longest planned path and generates the dates for activities based on their durations and logical sequencing.

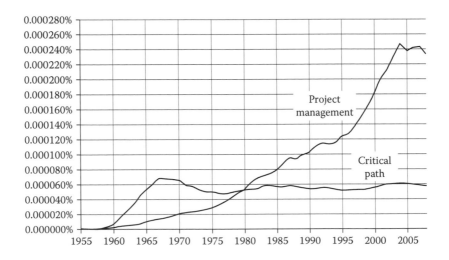

Figure 4.1 Google Books Ngram Viewer graph for "critical path" and "project management."

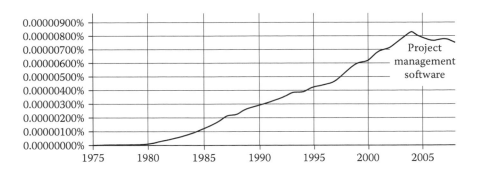

Figure 4.2 Google Books Ngram Viewer graph for "project management software."

The anomaly is that while usage of the term "project management" became far more widespread once software packages that enabled the discipline became readily available, usage of the basic functionality of the software, that is, critical path calculation, stagnated. As computer and software prices declined, users who had little sense of how to use the array of supportive functionalities started buying cheap packages that they would then use to "draw pretty pictures." The printed Gantt chart (a schedule display technique developed in the early 1900s by Henry Lawrence Gantt) became ubiquitous, but reflected dates more likely to have been determined by whim than by CPM algorithm.

The lack of maturity among project management software users reached the point where it was driving the technical support personnel of project management software companies crazy.

"I just bought your software," a caller would say, "but when I tried to use it, it started highlighting some of the activities in red. I don't know why it's doing that. I want it to stop."

"Sir, those activities are your critical path. As soon as you start entering activities, our software automatically calculates the critical path for you."

"How does it know what's critical to me? Several of these activities aren't critical to me at all! I want it to stop doing that. I know what's critical and what's not."

This may have been a suitable opportunity to conduct training on critical path theory, but such is not the function of technical support. In 2003, one leading software maker abandoned the practice of automatically highlighting the critical path. Thereafter, the user who knew enough to want the critical path highlighted had to pull down a "wizard" menu and instruct the software how to highlight it. With the curiosity that may have been aroused by apparently random activities suddenly turning red no longer a factor, knowledge of critical path analysis has undoubtedly atrophied even more.

Activity identification and duration estimating

This chapter is not intended to be a comprehensive text on critical path scheduling. Readers desiring such information should read my 1999 book, *Total Project Control: A Manager's Guide to Project Planning, Measuring, and Tracking.* However, an understanding of the basics of critical path analysis (and especially the new metrics critical path drag, drag cost, and true cost) is essential to the management of projects as investments: time is money, and the critical path determines project time.

In order to create a critical path schedule, we must first start by identifying all the work activities we intend to do. The best way to accomplish this is through construction of a work breakdown structure (WBS; again, covered comprehensively in my previous book). The WBS can be decomposed to provide us with a list of all the work activities.

The next step should be to estimate the *duration* of each of those activities, that is, the elapsed time from each activity's start to its finish. Note that duration is measured in hours, days, and weeks and not in work-hours (or person-hours), workdays, and workweeks. The latter units measure what is called *effort*. Duration is a function (often a complex one) of effort and applied resources.

Estimating duration almost always requires assumptions about effort and resources. My recommendation is always to start by assuming a minimum of one dedicated resource of each required type for each activity if assigning less would change its duration estimate. Why? Because if we can get 25% of a resource, we can almost always get 100% of that resource; we just need to be able to demonstrate the difference between what we could accomplish (in expected project profit terms) with a dedicated resource versus our limitations due to the restricted resource availability.

The above paragraph may seem simple, but it actually is of tremendous importance to project-driven organizations, whose ability to establish satisfactory staffing levels is rendered impossible by the arbitrary nature of project managers' assumptions about resource availability when doing activity duration estimates. When the initial estimates are made, no one knows which activities will be on the critical path (nor should they, as jumping to those conclusions can often result in the sort of wishful thinking that distorts duration estimates).

When estimating durations, a project manager will often think to herself, "If I could get Henry assigned to this activity full time, it would only take 4 days. But Henry's so busy that nobody ever gets him more than 25% of the time, so I'd better estimate it to take 16 days." The fact that this activity's duration is so resource dependent means it is what is called a resource-elastic activity.

This project manager's approach leads to five results, all of them bad:

1. The project manager is never able to demonstrate how much more valuable her project would be if she had a dedicated Henry (and if Henry's activity winds up on the critical path).
2. She therefore never gets a dedicated Henry, because she can never justify it.
3. The project therefore takes longer than it should and generates less project profit. But because the resource delays (caused by the part-time Henry) are conflated with the CPM delays caused by the nature of the work and the logical order in which it must be performed, this loss is never tied to the specific resource shortage that causes it.
4. Henry's functional manager (and the entire organization) is never able to demonstrate, in clear and monetized terms, how much it's costing the organization on all projects to be limited to only one Henry rather than hiring or training one or two or three more.
5. Staffing thus becomes shooting in the dark, with no one ever able to tell what the truly critical resources are.

Of course, if the project manager is really not going to be able to get Henry full time, then her project plan must ultimately reflect that and be adjusted accordingly. But such a decision should be made much later in the planning process, when resource scheduling is done on the basis of resource availability as reflected in the organization's resource library (and as we discuss in Chapter 7). But every organization's planning procedures should stipulate assembly of the initial CPM schedule on the basis of an estimating requirement of "a minimum of one dedicated resource for resource-elastic activities."

Hard and soft dependencies

Having identified the work activities and estimated their durations, the next step in CPM scheduling is to "sequence" the activities, that is, arrange them in the logical order in which they must be performed. This is done by identifying the *predecessors* or *dependencies* of each activity, in other words, determining what work must come immediately before other work.

Schedulers divide dependencies into two types:

1. A *hard* dependency is one that is based on the logic of the work: we can't paint the wall until we've plastered, nor manufacture the widget until we've received the raw material, nor debug the software until we've written the code.

2. A *soft*, or *discretionary*, dependency is one that is not driven by the logic of the work but by some other consideration such as, "As long as we've got the equipment here, let's do everything at once even though it will delay us a bit."

Frequently, soft dependencies are entered into the software to make sure that activities requiring a single resource are scheduled to occur serially rather than simultaneously. Once again the problem becomes one of conflation: it is very difficult (often impossible) to get around a hard dependency. But it is usually possible to mitigate a soft dependency, often by simply acquiring additional resources. If the cost of a soft dependency in schedule delay and reduced project value is substantially greater than the cost of the mitigating resources, then those resources should be acquired. But if soft dependencies are mixed in with hard dependencies, it quickly becomes impossible to tell which are which and what their impacts are.

As a result, in assembling the initial CPM schedule, my recommendation is to avoid all soft dependencies. Then by introducing each proposed soft dependency one at a time, it is possible to see the impact of each and identify where the cost of such an impact is unacceptable.

Basics of the CPM algorithm

We arrange the work activities into the sequence mandated by the hard dependencies, never starting a successor activity until all of the work that must come before it has been finished. We can then display this information as a flowchart. Such a chart for a very small project is shown in Figure 4.3. The diagram shows that Activities B, C, and D cannot start

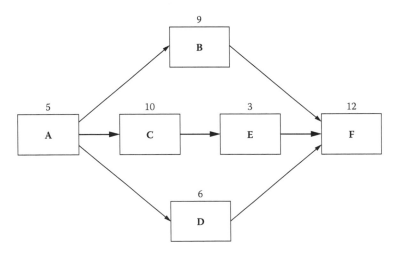

Figure 4.3 A simple flowchart for a project with six activities.

until Activity A is finished, E cannot start until C is finished, and F cannot start until B, E, and D are all finished.

If we are using project management software, the standard CPM algorithm is now implemented to generate a schedule for every activity in the project, whether there are 6 activities or 30,000. If we don't have such software, then we should apply exactly the same algorithm by using our brains rather than a computer. This can actually be a lot of fun, not unlike a Sudoku puzzle, but admittedly it is much easier with 6 or 60 activities than with 30,000!

The CPM algorithm actually consists of two separate algorithms. The *forward pass* algorithm traces the dependencies and durations from first activity to last and determines the earliest that each activity can start (ES) or finish (EF). The *backward pass* algorithm then goes to the end of the last activity and traces the network logic backwards to determine the latest that each activity can finish (LF) or start (LS) without delaying the last activity beyond its early finish date. By convention, this information for each activity is usually displayed in a box such as that shown in Figure 4.4, with the early dates on top and the latest possible dates on the bottom.

This sort of schedule flowchart is called a network logic diagram (or sometimes a PERT chart, due to anachronistic terminology). The network logic diagram for our six-activity project would look as shown in Figure 4.5.

The CPM algorithm has now told us:

- What the project duration will be if it is performed in this manner
- Which activities comprise the longest, or critical, path that determines the project duration
- What the earliest and latest possible dates are for each activity

The algorithm also quantifies the amount of time that activities not on the critical path can slip without delaying the end of the project. This is called an activity's *total float* (TF) or in some software packages, *total slack* (TS). It is the difference between an activity's late finish and its early finish (TF = LF – EF). Figure 4.6 displays the total float calculations below the box of each activity. The total float of Activity B is 4 and of Activity D is 7. But it is the total float of the critical path activities that

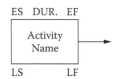

Figure 4.4 Format for an activity box in a network logic diagram.

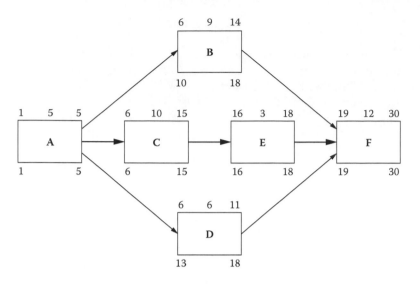

Figure 4.5 A CPM schedule in network logic diagram format.

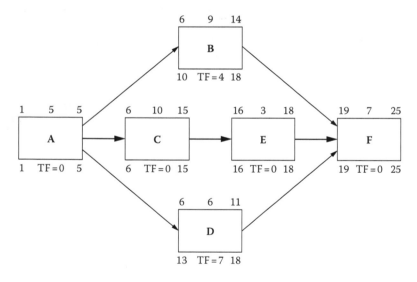

Figure 4.6 A CPM network logic diagram with total float computations.

is most striking: zero in every case. If we think about it for a second, it makes sense: the critical path activities determined the project duration, so if any one of them slips at all, the project completion will be delayed.

All project management software packages compute total float. Most also compute something called *free float* (FF), an interesting-to-know measurement of how much time an activity can slip without delaying any other activity. But notice: both total float and free float only occur on noncritical activities. If an activity is on the critical path, what does the software tell us about it? Answer: zero. It tells us that its total float is zero, which is the same thing as just repeating back to us that it is on the critical path.

Okay, so here is the trick question. What is more critical: information about something that is critical or about something that is not critical? There is an old saying that what is measured is what you pay attention to. In critical path analysis, what is measured is only what is off the critical path. As shown in Figure 4.7, all the attention is drawn to Paths A, C, and D. Path B, which determines the duration of the project and thus has a huge impact on the value of our project investment, is unquantified.

The typical result is that schedulers start by taking it as a given that the project duration will be equal to the length of Path B: 110 units. Then they look at the other item of information the software is computing, the quantification on the paths with float. And they conclude: "Huh! It really doesn't matter if Path A takes 35 longer, or Path C takes 20 longer, or Path D takes 25 longer; the project will still have a duration of 110 because that's how long Path B is." There is even a name for this approach. It is called *pacing the project*, working the noncritical paths at a speed where their durations will expand equal to the duration of the longest path. This approach is completely backwards. Instead, the attitude should be: "How can I compress Path B and thus make my project more profitable?" The problem stems from the fact that CPM tells us zero about the critical path activities, that the total float is zero.

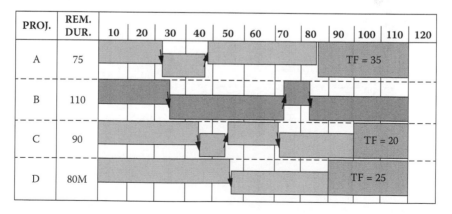

Figure 4.7 A Gantt chart showing a critical path and total float on noncritical paths.

Critical path drag

What should the software tell us? It should tell us information that will help us increase our expected project profit. As we discussed earlier, schedule compression will almost always increase our EPP. Schedule duration is driven by the critical path, about which the software tells us zero. What information do we need in order to help us identify opportunities for shortening the critical path? We need to know how much time each activity on the critical path is adding to the project duration so that we can see how much time we could potentially save by shortening it.

On a real project, we may have hundreds or even thousands of activities on the critical path, each involving different types of work, different types of resources, and different opportunities for compression. Ultimately, we will need to address each activity differently depending upon the individuality of its work. But for the moment we are simply looking at a very large diagram. Where do we start?

Let us return to the six-activity network logic diagram shown in Figure 4.8. How much time is each activity adding to the project duration or, alternatively, how much would the end of the project come in by if we were to shrink any one activity's duration to zero?

The first thing to recognize is that the two activities for which the software has provided us with some nice quantification are the two that are adding absolutely no time to the project duration. Compressing Activities B and D will do nothing to shorten the duration of the project because neither is on the critical path. To shorten the project we need to change

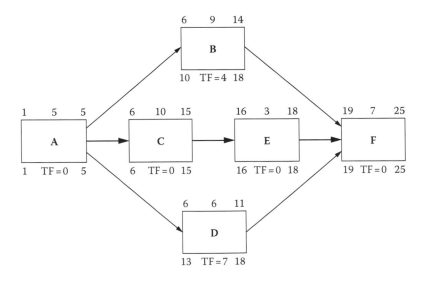

Figure 4.8 The CPM network logic diagram with float totals from Figure 4.6.

something about the way we are planning to do Activities A, C, E, and F, the ones that have that "TF = 0" under them.

Let's start with Activity A. Suppose we figured out a way to accomplish A instantaneously: instead of it having a duration of 5 days (or weeks; it makes no difference provided all the durations are in the same time units), what if we shortened it to zero days? How much would the end of the project be pulled in by?

The answer is easy to see. With Activity A taking no time, Activities B, C, and D could all start as soon as the project starts, at the beginning of Day 1. The result would be that the project would finish at the end of Day 20 instead of Day 25. Activity A was adding 5 days to the project duration, so that we can shorten the project by up to 5 days by figuring out a way to compress Activity A to zero duration.

What about Activity C? It starts with a nice fat duration of 10 days, double that of A. Suppose we were to figure out a way to compress Activity C's duration to zero? How much would that shorten the project?

The key is to recognize that, unlike Activity A, Activity C has two other parallel activities that are successors of Activity A and predecessors of Activity F: Activity B with a total float of 4 days and Activity D with a total float of 7 days. Now as we start to eliminate days from Activity C's duration, the total floats of Activities B and D will decrease on a one-to-one basis. If we succeed in compressing Activity C's duration to 6 days, the project will have been shortened by 4 days, down to 21 days. Activity D will still have 3 days of total float, but Activity C's total float will be down to zero, meaning that it is now also on the critical path. Any further time by which we now shorten Activity B will no longer shorten the project duration because the critical path will now go through Activity B. Any additional time above 4 days by which we shorten Activity C's duration will simply build up total float on Activity C and drag on Activity B. Activity C therefore has drag equal to the total float of the parallel activity that has the least total float, in this case, 4 days of drag, as shown in Figure 4.9.

Our next critical path activity is Activity E which, like Activity C, is parallel to Activities B and D. At first glance, therefore, one might assume that Activity E is adding the same amount of drag as Activity C, and that we can compress Activity E by up to 4 days before the critical path switches to Activity B. However, Activity E only has a duration of 3 days. Activity E is only adding 3 days to our project duration, and therefore the most we can compress the project by shortening Activity E is 3 days. Activity E's drag is therefore 3 days, as shown in Figure 4.10, despite being parallel with activities whose lowest total float is 4 days.

This refines the formula for drag calculation slightly: the drag of an activity that has something else in parallel is whichever is less: its duration or the total float of the parallel activity that has the least total float.

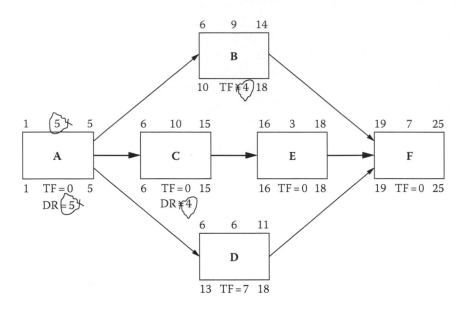

Figure 4.9 The CPM network logic diagram with drag computed for activities A and C.

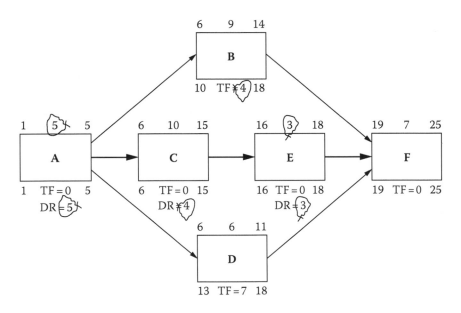

Figure 4.10 The CPM network logic diagram with drag computed for activities A, C, and E.

Finally, we have our last critical path activity, Activity F. Like Activity A, it has nothing in parallel (everything is a *descendant* of Activity A and an *ancestor* of Activity F) and therefore, like Activity A, its drag is equal to its duration: 7 days in the case of Activity F. See Figure 4.11.

The rules for computing critical path drag in a network where all the schedule dependencies are finish-to-start (FS), meaning that no activity can start until all its predecessors are finished, are

1. If a critical path activity has nothing else in parallel (i.e., everything else is either an ancestor or descendant), its drag is equal to its duration.
2. If a critical path activity has other activities in parallel, its drag is whichever is less: its duration or the total float of the parallel activity that has the least total float.

As simple and obvious as this concept is, it is almost unknown in project management. I discovered it in the early 1990s when as a consultant I started helping project teams compress their schedules, sometimes during the planning stage and sometimes during implementation when delays were making the schedule slip and the team was looking for ways to recover. My initial reaction was that this concept must have been widely understood, even though none of the software with which I was familiar seemed to compute it. It was only after I had searched through a multitude of books, articles, and software packages, that I realized I had,

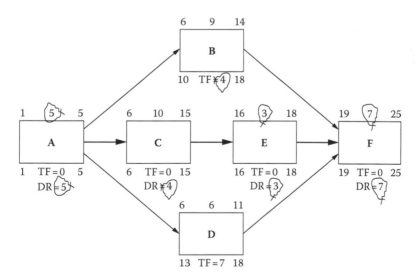

Figure 4.11 The CPM network logic diagram with drag computed for all CP activities.

as with the DIPP, stumbled across a new project management concept. And so I named it critical path drag,[2] drag being the force that slows down an object traveling through air or water.

Over the years, I have taught the concept in numerous graduate classes and PMI-approved seminars and webinars, published a book and numerous articles on the subject, and used it to help clients compress and recover schedules. There are now three software packages that compute critical path drag. I don't think there can be any doubt but that identifying how much time activities are adding to the project duration is a vital metric. And yet most of the project management world remains blissfully ignorant of the concept and fails to use it. Although the computations may seem simple here, that is due to the fact that (1) we are dealing with a very small network and (2) all the relationships are finish-to-start. On project networks with many hundreds or thousands of activities, it becomes difficult just to identify all the parallel activities. And once we introduce what are called *complex dependencies* (i.e., start-to-start, finish-to-finish, and start-to-finish relationships, abbreviated as SS, FF, and SF, respectively) and *lags* (delays between any two events), drag can be difficult to compute in even a relatively small network. In fact, it is exactly the kind of arithmetic metric for which computer software can be easily designed. The fact that most software does not compute it is a shortcoming that perhaps this book will help to correct.

The audience for whom this book is intended, executives and managers at the senior, functional, and project levels, need to be aware of drag and its implications. They also need to be aware that this metric is new and that their project managers, team members, and schedulers may be completely unaware of it. Senior managers should ensure that their projects are using the latest techniques, and should reinforce this by asking at every progress meeting to see a report listing the remaining critical path activities in descending order of their drags. After that, a few questions such as, "What have you done to look into reducing the top three or four drags?" are needed.

Managing change

There is a little joke in the project management world: it's in response to the question, "Why should we spend so much time planning the project?" The answer: "Because when we have finished identifying all the work activities and scheduling them, and scheduling the resources and developing the budgets and cash flow and risk plan and earned value baseline, we will have eliminated one of the millions of possible ways in which the project could actually go!"

Although it's a joke, there is an important element of truth to this. If anyone thinks that, because they have invested lots of time and energy in planning a complex project, it's actually going to go just as planned, they need to get out of the project business now! Small and simple projects may sometimes go as planned; large and complex projects never do.

So why do we bother to plan projects? The reason is because we know that things are going to change during execution. All of the primary tools and formats of project management have specifically been designed to be flexible, to reflect the changes as they occur, provide a framework to analyze the implications, and, with minimal effort and disruption, amend the plan to reflect the new realities.

As mentioned above, senior management does not need to understand the detailed calculations of critical path drag, just that the project managers are completely familiar with it and its uses. What follows is another fairly simple schedule, with answers, on which those who want to can practice their skills. Others can jump ahead to Chapter 5 where we focus on two key implications of critical path drag for the project investment: drag cost and the true cost of project work.

Riding a changing schedule

In Figure 4.12, we have a 10-activity network schedule in which the total project duration, early dates, late dates, and total float have been calculated and the critical path (A – B – F – I – J) has been identified. (In other words, it

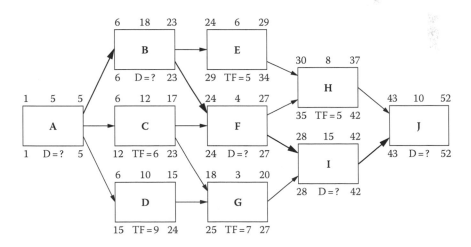

Figure 4.12 A 10-activity network logic diagram with drag not computed.

includes the data one would normally get from standard CPM scheduling software.) If you don't feel like giving yourself the exquisite fun of computing drag, just jump ahead to Figure 4.13.

1. *What is the drag of each critical path activity?* (This time let us assume the durations are in weeks rather than days; it makes no difference to the calculations, but 52 weeks seems like a nice number.) Take a look at the diagram and try to calculate the drags of the five critical path activities. Then look at the following answers and explanations.

Answers and explanations
a. *Activity A and Activity J:* These two activities, both on the critical path, have nothing in parallel so each will have drag equal to its duration: 5 weeks for A and 10 weeks for J.
b. *Activity B:* B has a duration of 18 weeks. It is a descendant of A and an ancestor of E, F, H, I, and J. By definition, it is parallel with everything that is not either an ancestor or a descendant. It is therefore parallel with Activities C, D, and G with total floats of 6, 9, and 7 weeks, respectively. Of those three, the lowest total float is 6 weeks on C, so Activity B's drag is 6 weeks.
c. *Activity F:* F has a duration of 4 weeks, is a descendant of A, B, and C and an ancestor of H, I, and J. It is therefore parallel with Activities D, E, and G which have total floats of 9, 5, and 7 weeks, respectively. So Activity F's drag *would* be 5 weeks except that its duration is only 4 weeks. So its drag is only 4 weeks.
d. *Activity I:* I has a duration of 15 weeks, is a descendant of Activities A, B, C, D, F, and G and an ancestor of J. It is therefore parallel with E and H, each of which has a total float of 5 weeks. So Activity I has drag of 5 weeks.

The answers are shown in Figure 4.13.

With drag computed, the key now is to understand that whatever changes occur to our schedule, we will be able to see the impact, measure it, determine other impacts, and potentially take other steps to lessen any damage. Let's make the sort of change that might occur during project implementation and see what the impacts would be. It's probably more likely to be a schedule slippage, but let's be optimistic and see what happens if an activity finishes earlier than planned; the same methods would apply for a slippage.

2. *What if Activity B takes only 10 weeks instead of 18?*
a. *How much would the end of the project come in by?* Because Activity B has only 6 weeks of drag, the end of the project would be pulled in

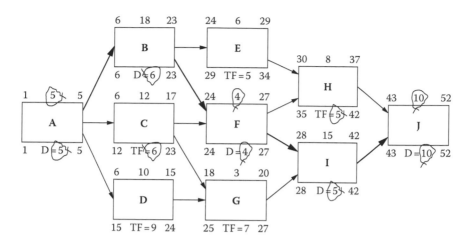

Figure 4.13 A 10-activity network logic diagram with drag computed.

by 6 weeks for a total duration of 46 weeks instead of 52 weeks. Even if you were to shorten Activity B to zero, the end of the project would still only come in by 6 weeks because B is only adding 6 weeks to the project duration. On any single critical path activity, you cannot shorten the project by more than that activity's drag. Now the critical path will go through Activity C instead of Activity B.

b. *What would Activity B now have (with a 10-week duration) instead of 6 weeks of drag?* The first 6 weeks removed from Activity B's duration would all be drag, at the end of which B would have zero float and zero drag. Shorten it by 2 more weeks and it would have 2 weeks of total float.

c. *What would Activity C now have instead of a total float of 6 if Activity B's duration is only 10 weeks?* With each of the first 6 weeks that's removed from Activity B, Activity C's total float would decrease by one week. When Activity B's duration is 12 weeks, both Activity B and Activity C will have zero float and zero drag, and we will have parallel critical paths. Take 2 more weeks out of Activity B and the critical path will change to go through Activity C alone, which will now be parallel with Activity B's 2 weeks of float and will therefore have 2 weeks of drag.

d. *What would Activity D now have instead of a total float of 9?* Activity D used to have 3 more weeks of total float than Activity C. If C is now the critical path with zero float, Activity D will still have 3 more days of total float than C and G, or 3 days total float in all.

The results of these changes are shown in Figure 4.14.

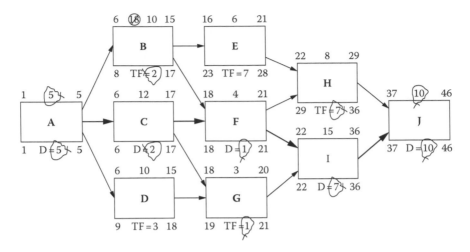

Figure 4.14 A 10-activity network logic diagram with computations after adjustments.

This book is not intended as a text for master schedulers or project management software designers, who are the people for whom calculating drag in networks with complex dependencies would be important. Any such readers should read my previous book *Total Project Control*, where I spend almost a hundred pages on calculating drag with all kinds of dependencies. But the above examples should provide an understanding of how computing drag will enable schedule compression and optimization.

A tragic example from history

Everything isn't always about money. As an illustration I offer this story that I included in a sidebar to an article "The Drag Efficient" that I published in the January–February 2012 edition of *Defense AT&L Magazine* (Figure 4.15). In war, a single day of delay can have disastrous impacts. For the American soldiers at Dhahran, the cost of one day was 28 lives and the destruction of 28 families. In other wartime scenarios, the cost of delay could be a lost war.[3]

But bad project management causes lost lives in a wide range of disciplines. Businesses, government departments, and international aid organizations engaged in trying to save lives need to start recognizing that poor project management is actually costing lives. The people who work in such industries are conscientious and often make great personal sacrifices. I suspect that they would be horrified to know that their ignorance of PM methods often cost a percentage of the human lives they are working so hard to save.

**A HISTORICAL EXAMPLE OF DRAG
COST IN HUMAN LIVES**

In 1991, during the first Gulf War, it was discovered that a software bug in the radar of the Patriot antimissile system was causing the timing system to lose a small fraction of a second for every hour that a battery had been operational. Quoting from the February 4, 1992, report of the Information Management and Technology Division of the United States General Accounting Office (http://www.fas.org/spp/starwars/gao/im92026.htm):

> On February 21, 1991, the Patriot Project Office sent a message to Patriot users stating that very long run times could cause a shift in the range gate, resulting in the target being offset. The message also said a software change was being sent that would improve the system's targeting. However, the message did not specify what constitutes very long run times...

> ...Alpha Battery, the battery in question, was to protect the Dhahran Air Base. On February 25, Alpha Battery had been in operation for over 100 consecutive hours. Because the system had been on so long, the resulting inaccuracy in the time calculation caused the range gate to shift so much that the system could not track the incoming Scud. Consequently, Alpha Battery did not engage the Scud, which then struck an Army barracks and killed 28 American soldiers. On February 26, the next day, the modified software, which compensated for the inaccurate time calculation, arrived in Dhahran. According to Army officials, the delay in distributing the software from the United States to all Patriot locations was due to the time it took to arrange for air and ground transportation in a wartime environment.

Although there is always a strong tendency to blame the last few activities (i.e., "the time it took to arrange for air and ground transportation") for a late delivery, the fact is that every critical path activity contributes to the project's duration. In this case, every activity that had drag of 1 day or more, and that might somehow have been shortened through additional resources or expense, could have saved the lives of those 28 soldiers.

Figure 4.15 Sidebar about loss of life from my *Defense AT&L Magazine* article, January–February 2012.

Summary points

1. Although in recent decades project management has become fully recognized as a business discipline, knowledge of one of its crucial techniques, critical path analysis and scheduling, has atrophied.
2. Every project has a critical path, whether the project team chooses to plan and manage it or not.
3. The longest path of activities, constraints, bottlenecks, and delays always determines the length of the project. It is therefore qualitatively different from every other path, and anything that delays the critical path can have a major impact on the project's business value as it will delay project completion.
4. Traditional critical path theory quantifies float (slack), which is the amount by which an activity is not on the critical path. Only noncritical activities have float. Much more important is the amount of time that an activity is adding to the project duration, as that can affect both the project's value and its cost. This is called *critical path drag*. But neither traditional critical path theory nor the vast majority of project management software packages compute this crucial metric.
5. Critical path drag computation allows the project manager to see which activities are delaying the project the most and to make changes to compress the schedule.
6. The project will almost never go precisely as planned. Changes to the schedule can be expected to occur on every project. A critical path schedule allows the project manager to see the implications of any change in any part of the schedule.

Endnotes

1. https://books.google.com/ngrams/graph?content=project+management+software&year_start=1975&year_end=2008&corpus=15&smoothing=3&share=&direct_url=t1%3B%2Cproject%20management%20software%3B%2Cc0
2. Initially I made it an acronym—DRAG, all in caps—which stood for Devaux's removed activity gauge. But eventually I came to realize that if I wanted project managers to appreciate its value, I needed to tie its name more clearly to the critical path. "Critical path drag" seems to have caught on.
3. We will admit that sometimes this can be a blessing. Adolf Hitler's decision in the spring of 1941 to support Mussolini's troops in Greece meant that he had to delay the German invasion of Russia by several weeks. The added drag meant that the Wehrmacht offensive was slowed by autumn mud and the onset of winter. This left his panzers a few kilometers short of Moscow and gave the Russian Army time to regroup and ultimately win the war.

chapter five

Optimizing the schedule with drag and drag cost

> "Where can you go to buy time, and how much should you pay?"

In a recent online conversation about critical path drag with an experienced and competent project manager, I suddenly discovered that despite his knowing about the concept, he really didn't appreciate how to use it. I therefore explain in this chapter how to use drag both to optimize a schedule during upfront planning and to recover a schedule that has slipped.

Unfortunately, optimizing the schedule during the planning phase is something that is rarely done, largely because project managers don't know how to do it. They often don't even understand exactly what the term means. Insofar as they think about improving the schedule, they think about either making it more dependable and stable, or making it shorter. But although either of those factors may be desirable, that is not what I mean by an optimized schedule. If a project is an investment, then an optimized schedule is one that will give our project a higher expected project profit and usually a higher DIPP.

Drag computation in combination with an understanding of the cost of time, both acceleration premium and delay cost as we discussed in Chapter 3, is what allows us to optimize the schedule for our project investment. And when we make a modification that genuinely improves our schedule (taking into account risk factors), we can tell because our project EPP should be higher.

Drag cost

In certain ways, the most important implication of drag is only unearthed by relating it back to our discussion in Chapter 3: the value/cost of time as an acceleration or delay of project completion. The fact that an activity has drag of 7 days and therefore delays the end of the project by that amount of time is interesting enough, but the fact that those 7 days are reducing the expected project profit by $20,000 each, or a total of $140,000, is the kind of information that not only makes managers sit up and take notice; it's also the kind of data that can be used by the project team to increase the EPP and the DIPP. We are not now simply saying, "Let's make this project shorter and it will be worth more." Instead, we are saying, "The duration of this specific activity is costing us 12 days times $20,000, and this one is costing us 8 days times $20,000, and this one is costing us 5 days times $20,000." The information is not only monetized, it is directed at specific work activities that can now be explored for ways to compress them, perhaps even by spending more on resources to reduce a greater amount of drag cost.

Let's assume that we are performing the six-activity project from Chapter 4 for a customer. (The network diagram is reprinted in Figure 5.1.) Let us assume that the customer really needs the project to be completed in 24 days; every day later is going to cost him a lot of money. The fixed price contract therefore includes a penalty for late delivery. If we complete it in 24 days, we will be paid $400,000. But every day later will be subject to a penalty of $20,000.

We have computed the drag for the critical path activities and now know exactly where to look if we want to shorten the project. But is there any value in doing so? If this is the schedule we execute, the best we can expect to do is 25 days, which will result in a delay penalty of $20,000 and total payment of $380,000. Therefore it seems as though there is value in eliminating 1 day of drag from any of the 4 critical path activities.

But there is a vital item of information that we don't have, and which contractors frequently do not have because they rarely bother to ask

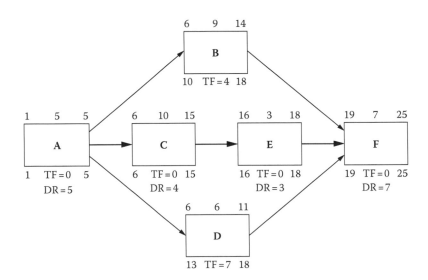

Figure 5.1 Network logic diagram of the six-activity project from Chapter 4.

the customer: is there any value to the customer of finishing the project earlier? Instead, the project manager and team settle for:

- A schedule that will have a duration of 25 days, thus generating $380,000 and leaving $20,000 on the table.
- Compress the schedule to a duration of 24 days, and hope that nothing unexpected will happen and they get the full $400,000.
- If the project manager really knows what she is doing, compress the schedule to 21 or 22 days, leaving the extra time as schedule reserve in case that unexpected thing actually does happen. But when she nears the end of the project and still has 2 or 3 days of schedule reserve in hand, she loses any sense of urgency and delivers the product to the customer at the end of Day 24.

Schedule reserve is time, and time almost always has value. The chances are that if there is going to be substantial cost to the customer if the project extends beyond Day 24, there is also value if the project finishes earlier.

- At the very least, finishing the project earlier eliminates the risk of finishing late. A time-driven customer will often be willing to pay an early delivery premium just for that peace of mind.
- The customer might be able to start using the deliverable earlier, in which case the reduced time will almost certainly have a lot of value.
- The customer might be dealing with a market window (such as a seasonal market) or be in a race with a competitor, in which case finishing earlier could have a great deal of added value.

But if there is value to the customer for the contractor to finish the project earlier, wouldn't the customer have so stipulated in the contract? Not necessarily, because customers usually don't fully appreciate the nature of a project either. Large contractors frequently subcontract portions of their project, and although one would think they would understand enough to write a contract that maximizes the benefit they will receive, it's been my experience that they do not know how to write those contracts either. Everyone is in thrall to the concept of the deadline, ignoring the very real damage that Parkinson's law and the principal–agent problem can generate.

Having developed the initial schedule that shows a project duration of 25 days, the project manager should do two things:

1. Instruct the project team to start looking for opportunities to pull the schedule in as much as possible even if it means increasing cost somewhat.
2. Meet with the customer to find out if an earlier delivery date would have any value for him.

It is astonishing how often project managers look upon the terms of a project charter or contract as being set in concrete. Any contract can be amended by agreement of all the parties. The sponsor or customer typically has no idea about the detailed planning of the project, and thus no idea about opportunities to do the work differently that the project team might discover. If alternative ways of performing the project might change for the better any of the parameters of scope, cost, time, and risk and might give the sponsor/customer greater benefit, the project manager should disclose this. She almost certainly has a vested interest in doing so, as this may represent a valuable business opportunity to increase value to both customer and contractor, and to be the project manager who identified a way of bringing in additional revenue on the contract.

Let us presume that in talking with the customer, the project manager discovers that, even though the current contract calls for a 24-day delivery, the customer would love to take delivery earlier. She should point out that, although it may be possible to accomplish this, it is likely to necessitate greater cost: resources may have to be added and employees may need to be paid overtime or given incentives to work long hours.

"I'm not absolutely certain that we can deliver earlier," she should say. "But just suppose we could find a way. What would it be worth to you for every day earlier we could deliver? Remember, if we can't deliver earlier, then you don't pay us anything extra; we only get the incentive if we deliver early."

This is how one can make a contract into a true win–win: aligning customer benefit with contractor incentives. If early delivery truly has value for the customer (and it does far, far more often than the infrequency of such incentive clauses would suggest), he will almost certainly agree to a mutually satisfactory amendment to the contract with an early delivery incentive.

Now the project manager can return to the project team with exciting news: every day earlier than Day 24 that the project is finished will be worth an additional $10,000 to the organization. Now up to $9,999 can be used for every day that can be saved, and our organization will be one dollar better off for every day.

Of course, because we are in the business of making a profit, we would rather spend a lot less than $9,999, say, $99 or $999 for every day we save. And that is the way we optimize our schedule to increase our expected project profit.

In Figure 5.2, the chart shows the EMV of the project depending on the day of completion and the cost of drag for any critical path activity based on how much it decreases the project EMV by delaying project completion. As the table shows, the current plan calls for completion on Day 25. With a delay penalty of $20,000, the EMV will be $380,000. If we can pull in the completion date to the contractual date of Day 24, the EMV will increase by $20,000. For each day earlier than Day 24, the EMV will

Completion	EMV	Delay Cost/ Accel Gain	Cum. Drag Reduction	Cumulative Drag Cost
Day 25	$380,000	−$20,000	0 Days	Curr. Plan
Day 24	$400,000	$0.00	1 Day	1d = $20,000
Day 23	$410,000	$10,000.00	2 Days	2d = $30,000
Day 22	$420,000	$10,000.00	3 Days	3d = $40,000
Day 21	$430,000	$10,000.00	4 Days	4d = $50,000
Day 20	$440,000	$10,000.00	5 Days	5d = $60,000
Day 19	$450,000	$10,000.00	6 Days	6d = $70,000
Day 18	$460,000	$10,000.00	7 Days	7d = $80,000
Day 17	$470,000	$10,000.00	8 Days	8d = $90,000
Day 16	$480,000	$10,000.00	9 Days	9d = $100,000
Day 15	$490,000	$10,000.00	10 Days	10d = $110,000

Figure 5.2 Chart of delay cost and activity drag cost with acceleration premium of $10,000.

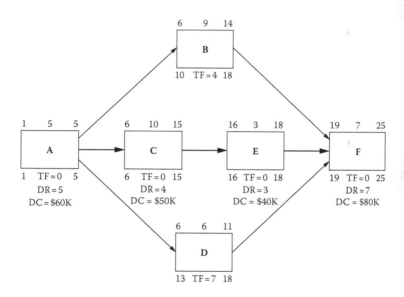

Figure 5.3 Network logic diagram displaying drag cost for each critical path activity.

increase by the acceleration premium of $10,000 per day. Looking at the Cumulative Drag Cost column on the far right, an activity with 1 day of drag will have a drag cost of $20,000, $30,000 for 2 days of drag, $40,000 for 3 days of drag, and so on. We can therefore plug the drag cost data into our network logic diagram in Figure 5.3.

Combining the data from the table with that from the network logic diagram, we can see that the critical path activities, due to their delaying the end of the project and thus reducing our EMV, have drag costs that range from $40,000 for Activity E's 3 days of drag to $80,000 for Activity F's 7 days of drag. The first day of drag on each activity pushes the completion to Day 25 and invokes the delay cost of $20,000. Every other day of drag on each activity costs $10,000 as it deprives us of a day's worth of acceleration premium.

Remember, this is all cost due to reduction in the project's EMV. It is also possible that, through the marching army costs factor due to project support activities, overhead, and opportunity costs that accumulate every day that the project goes on, compressing the schedule may also increase the project profit by reducing the cost of getting the whole project done. These costs should be analyzed and added to the drag cost calculation.

True cost of project activities

Now we come to a very important concept, but one that is basically unknown in finance and cost accounting departments. During the 25-plus years that I have been teaching corporate classes in project management, it has been extremely rare for me to have even a single finance or cost accounting person in my class. It seems that project management is considered a topic that relates only to project managers and project team members, even in organizations that are 100% project-driven (by which I mean that their entire revenues come from performing projects).

The truth is that in project-driven organizations, there is nothing more important than projects and that means there is almost nothing more important than critical path analysis. There may be employees who don't need to understand CPM, but their numbers are vanishingly small.

- Functional managers need to understand critical path data in order to prioritize the assignments of department resources to the projects where they are most urgently needed.
- People in the contracts department need to know if the subcontractor being hired is going to be doing work that is either on the critical path or has minimal total float. Critical and near-critical activities are very different from those with lots of float and the contracts for such subcontractors should reflect that difference.
- An HR specialist responsible for recruiting a new employee needs to understand the drag cost implications if that employee is urgently needed for critical path work, and either raise the starting salary or otherwise accelerate the recruiting cycle.

- An acquisitions specialist responsible for acquiring material or equipment needs to know whether they are a constraint on the critical path and expedite delivery if they are.
- And the entire finance department, if it is to accurately track costs and guide the organization in making cost-efficient decisions, must understand what the true cost is of critical path work activities.

The key thing is to recognize that work on the critical path is completely different from all other work. Low wage workers on the critical path can cost a project a lot more than high wage workers on activities with lots of float. And the reason is drag cost.

Let us return to our network diagram and this time assign a budget of $200,000 to the project, with individual budgets for the resources in each activity as shown in Figure 5.4: A casual glance at the table shows that Activity D and Activity B are the two most expensive activities at $60,000 and $50,000, respectively. However, if we take those same data and plug them into our project schedule, we may notice a different picture, as shown in Figure 5.5.

The resource budgets of Activities B and D are the total costs of doing the work in those two activities. But the drag costs of our critical path activities A, C, E, and F make it considerably more expensive to do the work in those activities. The true cost of doing work in a project is driven by

1. Whether it's on the critical path
2. If it is, how much drag it has
3. How much time it is adding to the project duration costing in terms of reduced expected project profit

We can now sum the resource cost of each activity with its drag cost to get its true cost, as shown in Figure 5.6. It now becomes clear that Activities B and D are not in fact the most expensive because, being off

Act.	Duration	Budget
A	5 Days	$20,000
B	9 Days	$50,000
C	10 Days	$30,000
D	6 Days	$60,000
E	3 Days	$15,000
F	7 Days	$25,000

Figure 5.4 Budgets for each activity.

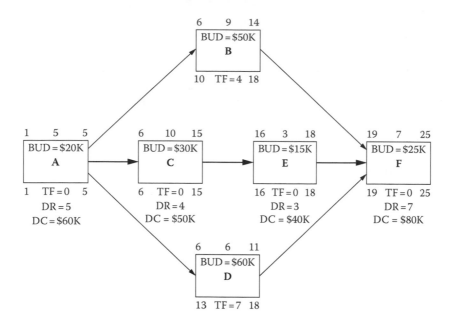

Figure 5.5 Network logic diagram displaying drag cost and budgets for each activity.

Act.	Duration	Budget	Drag Cost	True Cost
A	5 Days	$20,000	$60,000	$80,000
B	9 Days	$50,000	$0	$50,000
C	10 Days	$30,000	$50,000	$80,000
D	6 Days	$60,000	$0	$60,000
E	3 Days	$15,000	$40,000	$55,000
F	7 Days	$25,000	$80,000	$105,000

Figure 5.6 Summing budget and drag cost to get the true cost of each project activity.

the critical path, they have no drag cost. Activity B actually has the lowest true cost of all six activities and Activity D, despite having the largest resource budget, has a lower true cost than any of Activities F, A, and C.

The value of this information to a project manager should be clear. The eternal complaint, that "They won't give me the resources I need," is poignantly addressed by the drag cost and true cost concepts. If increasing the resource cost of an activity by a certain amount decreases its drag cost by even more, then the true cost of that activity will be less. The calculation is a simple one: trading resources for the value of time.

It is important to recognize that all the information we have thus far is based on estimates—perhaps estimates that have been informed by historical data mined from actuals on previous work—but estimates nonetheless. Given estimates that seem only slightly to support one decision over another, a reasonable sponsor might well be reluctant to spend more money on resources to get only minimal improvement in true cost.

For example, what if the sponsor is told that by adding $25,000 to the budget for Activity C, the duration will be shortened from 10 days to 8 days? The reduction of 2 days of drag on Activity C will compress the planned project duration to 23 days. This would eliminate the delay penalty of $20,000 as well as add one day of acceleration premium worth $10,000, for a total increase in the project's EMV of $30,000 and of the project's EPP of just $5,000 ($30,000 minus $25,000 for the resources). To spend an extra $25,000 of expense to gain such a small amount, even if the time gain were highly probable, may not be deemed worthwhile. The chances are that the time gain is anything but certain. And even if it is 80% probable, that 20% uncertainty factor is worth 20% of $30,000, or $6,000. That probability-weighted reduction in the equity resulting from the added expenditure is enough to advise against the additional investment: $25,000 minus (80% of $30,000) = $25,000 minus $24,000 = –$1,000. Spending the extra $25,000 would, on average, reduce the EPP by $1,000.

When a project team first starts using drag and drag cost to optimize a schedule, I have found that it usually unearths a multitude of opportunities where the potential rewards are five or ten times the required additional investment: an extra $20,000 of budget on specific activities to reduce drag cost by $100,000 and more. In such cases, the estimates can be off by wide margins and the results will still justify the additional investment. If an organization is cash-strapped, then it may not have the funds to increase a project budget; if there's no money, there's no money. But in any investment, if thorough risk analysis demonstrates that the return can be substantially augmented by an incremental increase in invested funds, it is almost always a good decision provided the funds are available.

Resource elasticity and the DRED

There is a feature of most work activities called *resource elasticity*. This is the property of an activity's duration to lengthen or compress in response to changes in the amount of assigned resources. This aspect of work activities has been well known for a long time, and planners will sometimes generate secondary duration estimates for work activities called the *crash duration*. The crash duration estimates the least amount of time that an activity will take even if given unlimited resources. For example, it may be estimated that Activity C, with an estimated duration of 10 days, could be shortened to 3 days, if almost unlimited resources were assigned to it.

Although the intent to have a measure of activities' resource elasticity is a good one, there are three problems with using crash duration:

1. Usually the crash duration estimate is modeled as a linear relationship between cost (or resources) and time, although such is usually not the case. There is usually a point of diminishing returns, where the incremental increase in resources is not matched by a similar increase in productivity. If a 10-mile ride takes one hour on a horse, but we estimate a crash duration of 30 minutes if we use a relay of four horses that we change every 2.5 miles, would we really compress the duration by 10 minutes for each additional horse: 50 minutes with two horses and 40 with three horses? Doesn't it seem more likely that two horses would make it in 45 minutes and three horses in about 35 minutes? Frequently the impact of each additional resource seems to obey absolutely no mathematical rules: it will take one person an hour to drag a long table up to the attic; two people can do it in 10 minutes; three people may actually take 12 minutes, getting in each other's way; and four people might be able to do it in 9 minutes, one person on each corner.

2. Who really understands what crash duration is, anyway? Are we really talking about the least amount of time something would take if we use the combined forces of NATO? Or is it just our internal resources? Or something in between? Whatever assumptions one estimator makes, another may make different ones.

3. Crash duration is just not a very practical estimate, in that we are very rarely going to be able to get anything close to the number of resources that would make such a duration possible.

A far more useful way of estimating resource elasticity is to set the amount of additional resources at a fixed quantity and estimate the impact of that increase on the activity duration. And the simplest fixed quantity is double. I call this the doubled resources estimated duration (DRED), and it prompts the information in terms that any subject matter expert can easily understand: if you were to double the amount of resources for Activity X that you have assumed in making your initial estimate, what would the duration of the activity become? Some possibilities are shown in Figure 5.7.

The initial estimate for the duration of Activity X is 20 days. There are about six different possibilities listed for the duration estimate if the resources (and budget) for Activity X are doubled:

- DRED Possibility 1 is similar to the example of getting the long table up to the attic: there is an optimal size for a team doing certain tasks. Carrying a long table upstairs is one such task; it's a whole lot simpler and faster with two people than with one. In such cases, doubling the resources doesn't just halve the duration. It compresses it to

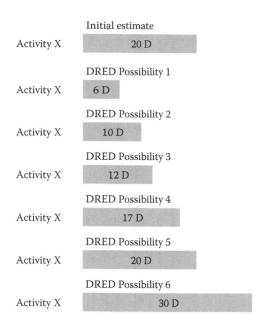

Figure 5.7 Some plausible DREDs for an activity with an estimate of 20 days.

less than half. It would be nice to think that project and functional managers always know when an activity has a resource estimate that is so inefficient, but such is not necessarily the case. Just prompting for a DRED estimate can correct such problems. Of course, if the activity is off the critical path and has lots of total float, then leaving one poor drudge to haul the long table up all those flights of stairs might be the cheapest and most project-efficient way to go, but it just seems offensively inefficient. Surely it usually makes sense to get the task finished faster by adding a second laborer and thus freeing them both up for other duties.

- DRED Possibility 2 is an example of an activity that is *perfectly resource elastic*. This means that doubling the resources halves the duration. All else being equal (which they almost never are), the budget for doing such an activity should be identical whether using the initial estimate or the DRED: double the resources for half the time should equal the cost using the original estimate (or even be less if we have reduced overhead costs and opportunity costs).
- DRED Possibility 3 is very common: we get a big compression of duration for doubling the resources, but not quite 50%. For activities on the critical path with substantial drag, such DREDs represent major opportunities to increase the expected project profit by adding resource cost to reduce drag cost, and thus also true cost.

- DRED Possibility 4 represents a work activity that is only slightly resource elastic. Perhaps the initial estimate is already based on an optimum team. However, if there is a small amount of compression to be gained, that might still be worthwhile especially if the activity has a high drag cost and a low initial budget. We talk more about this in a short while.
- DRED Possibility 5 shows an activity that is not resource elastic. It may sometimes be the case that simply doubling the resources does not bring us to an optimum team level. For instance, moving a baby grand piano upstairs may be as difficult and time-consuming for two people as it is for one, but if we quadrupled the resources to four, that might reduce the time. When faced with an activity whose DRED is identical with the initial estimate, it may be worth it for the project manager to ask why. Most of the time it probably means that the work is resource inelastic. This is valuable to know. We not only shouldn't waste extra resources on it, we should also check that the activity's resources are already at a minimum appropriate level.
- DRED Possibility 6 is the "too many cooks" example, and it is not as unusual as one may think. On many jobs, additional resources may actually slow down the work. Installing the avionics equipment in a jet fighter cockpit is one example: the engineers get in one another's way. Or sometimes the cooks may simply have different ideas of how to do things and the resulting disagreements require more time to be resolved than any gains from the additional resources. One caveat: in my experience, functional managers will often assess an activity as being Possibility 6 when, with proper analysis and division of labor it could be Possibility 4, offering small but possibly significant time gains. An understanding of the concepts of quantified drag and drag cost as attached to specific work activities can inspire functional managers and others to do that sort of analysis.

Using the DRED with true cost

In any financial enterprise, it is useful to know the precise nature of all the drivers of both value and cost. In the case of true cost, there are two components: resources and drag. The comparative size of each of these costs may provide an opportunity to trade one for the other. Those activities whose drag cost is a lot greater than their budget often represent such a chance to improve the expected project profit by increasing their resources while reducing their drag cost. The ratio of daily or weekly cost rate to drag cost therefore is an extremely helpful metric, and should be a standard report or column for every project management software package. In Figure 5.8, we see an example of such a report.

Act.	Duration	Budget	Daily Res. Cost Rate	Drag Cost	Daily Rate/ Drag Cost Ratio
A	5 Days	$20,000	$4,000	$60,000	6.7%
C	10 Days	$30,000	$3,000	$50,000	6.0%
E	3 Days	$15,000	$5,000	$40,000	12.5%
F	7 Days	$25,000	$3,571	$80,000	4.5%

Figure 5.8 Table showing each activity's ratio of daily cost rate to drag cost.

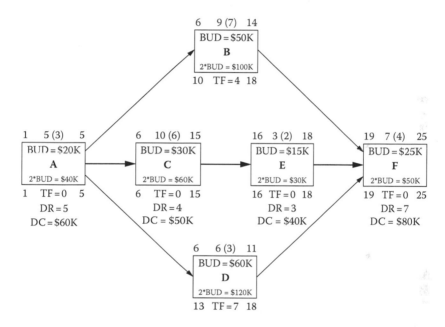

Figure 5.9 Network logic diagram with DREDs in parentheses and budgets doubled.

Any activity with drag and drag cost should be analyzed for opportunities to shorten the project and thus increase its EPP.[1] But on a large project, one may be staring at the network with hundreds of critical path activities. By initially focusing on those activities whose ratio of daily cost rate to drag cost is lowest, we are able to identify what may be the lowest hanging fruit, those activities where the resources to compress the activity duration and drag may be the least expensive.

That brings us to the other variable of the work activities: how comparatively resource elastic are they? And this is what the DRED measures. In Figure 5.9, we show the network logic diagram with the DRED duration estimate in parentheses. We should have collected this estimate for all activities (not knowing in advance which would be on the critical path and have drag) when we collected the data for our initial plan. Each activity's budget will be doubled if we go to the DRED estimate.

The DRED helps to point out with much greater precision the opportunities for investment gain through specific actions to compress the schedule. The table in Figure 5.10 makes the opportunities clear. As we can see, the impact of doubling the budget on Activity A from $20,000 to $40,000 will shorten its duration from 5 days to 3 days. Because A initially has 5 days of drag, the drag will be reduced to 3 days as the project duration will be compressed from 25 days to 23 days. The first day of reduced drag will reduce the delay cost by $20,000 for coming in from Day 25 to Day 24. The second day of reduced drag will further compress the schedule to Day 23 and add 1 day of early delivery premium worth an additional $10,000. So spending an additional $20,000 on Activity A reduces its drag cost by $30,000, resulting in a reduction in true cost of $10,000, from $80,000 to $70,000.

But it is crucial to understand that making any changes in one part of the network can cause changes in other parts of the network as well as in the outcomes of the project investment. If we change Activity A from 5 days to 3 days, the project duration will immediately be compressed to 23 days and the EPP will rise by $30,000. That means that compression on other activities will no longer generate quite as much drag cost savings as the first day of drag saved will no longer be worth the $20,000 delay cost. Is Activity A the best place to save that $20,000?

If we change Activity C's initial budget of $30,000 to its DRED budget of $60,000:

1. It will compress its duration from 10 days to 6 days.
2. It will completely erase its drag of 4 days.
3. It will pull in the project completion from Day 25 to Day 21.
4. It will add $50,000 to the EMV.
5. It will reduce Activity B's true cost by $20,000.
6. It will add $20,000 to the project EPP.

Act.	Initial Dur.	Dred Dur.	Initial Budget	Dred Budget	Initial Drag	Dred Drag	Initial Drag Cost	Dred Drag Cost	Initial True Cost	Dred True Cost	Improved True Cost
A	5 Days	3 Days	$20,000	$40,000	5 Days	3 Days	$60,000	$30,000	$80,000	$70,000	$10,000
C	10 Days	6 Days	$30,000	$60,000	4 Days	0 Days	$50,000	$0	$80,000	$60,000	$20,000
E	3 Days	2 Days	$15,000	$30,000	3 Days	2 Days	$40,000	$20,000	$55,000	$50,000	$5,000
F	7 Days	4 Days	$25,000	$50,000	7 Days	4 Days	$80,000	$40,000	$105,000	$90,000	$15,000

Figure 5.10 Table showing improved true costs due to use of DREDs.

In addition, the change in Activity C now would mean that Activity A's 2 days of drag only have a drag cost of $20,000, as the schedule is already into Day 21 where the delay cost of $20,000 is no longer at issue. Thus we can see that, if we can pull the schedule in elsewhere than on Activity A, there is no value in going to the DRED budget on Activity A; its drag cost reduction of $20,000 will be offset by its increased budget, and its true cost will remain unchanged.

If we change to the DRED budget on Activity C, it's crucial to be able to see what happens to Activity E, and the only way to visualize this is to look at the schedule. Suddenly, Activity B is also on the critical path. Its total float has disappeared, as has the drag and drag cost of Activity E. We now have parallel critical paths: A – C – E – F and A – B – F. In addition, the drag cost of both Activity A and Activity F is now simply $10,000 per day, with the $20,000 delay cost no longer in the picture. See Figure 5.11.

The question now is whether it is worthwhile to double the resources of Activity F to the $50,000 DRED budget. This would reduce its duration and drag from 7 days to 4 days, its drag cost by $30,000, and its true cost by just $30,000 minus $25,000 = $5,000. The numbers are close and one could go either way.

One more thing: the fact that we have parallel critical paths means that none of B, C, or E currently has drag. If we shorten any one of them,

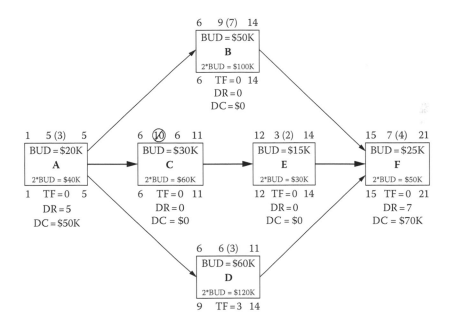

Figure 5.11 Adjusted schedule and drag costs after implementing the DRED on Activity C.

the end of the project will not be pulled in further unless we also shorten the parallel path. In many cases, it might be worthwhile to do so. However, Activity B has a large budget and a DRED that only allows for saving 2 days at a cost of $50,000. We have reached a point of diminishing returns.

Cautionary note on using the DRED

The DRED is an indicator and a tool; it is not magical or automatic. No project manager, having gotten estimates on duration and cost from activity managers or functional managers, should ever change them without first running the proposed changes by the original estimator (who is probably going to be the person who has to do the work!). This includes switching from the initial estimate of duration to the DRED estimate.

Instead, the project manager should use the DRED to analyze the plan, seeing where it might help and where it might not. Then the project manager must return to the estimator, show that person the implications of the estimates, and once again get input and assurances: will they really be able to reduce the duration from 10 days to 6 days if given double the resources?

In such cases, the project manager often receives a pleasant surprise: "Actually," says the activity manager, "I don't really need to double all the resources in order to save that time. If I could get a second truck (or physicist, or electrical engineer, or C++ programmer, or …), I wouldn't need any more laborers (or chemists, or mechanical engineers, or documentation writers, or …). So that would only raise the budget by 25%, not by 100%." Great! Let's do it!

In other cases, the project manager might find out that, yes, it would be possible to cut the duration in half if we could get another electrical engineer who specializes in this particular field of optics, but such people are scarcer than ice hockey goalies with their own front teeth. And even if we could find one, we'd have to pay such a hiring premium that it wouldn't be worth it. Back to the drawing board.

The DRED is a useful tool when combined with true cost. But it's not a magic wand.

Assessing optimization of the CPM schedule

So we have pulled in the schedule from 25 days to 21 days while increasing the budget from $200,000 to $230,000. The change in schedule has changed the project EMV from $380,000 to $430,000. Our original starting DIPP was $380,000 divided by $200,000, or 1.90. Our starting DIPP with the new schedule would be $430,000 divided by $230,000, or 1.87.

That is correct: our starting DIPP has gone down. But our expected project profit has gone up, by $20,000, from $180,000 to $200,000.

The reason that the DIPP has gone down even though the EPP has gone up is that the marginal return on the $30,000 it takes to pull in the schedule ($50,000 divided by $30,000 = 1.67) is less than the return on the original $200,000 investment (1.90). Should the sponsor/customer authorize the additional funds even though the DIPP, reflecting the return on investment per dollar, goes down? That is a judgment call. If there is a better way of investing the additional $30,000 than one which seems likely to return 1.67 for every dollar invested, then that would probably be the way to go. But a 67% return in a few weeks is not bad!

Of course, we can't be sure that the project will now finish at Day 21. Something may slip, reducing the acceleration premium and perhaps ultimately triggering an even larger delay cost. But that could happen anyway. What the compressed schedule has in effect done is to create a schedule reserve between the finish date of the working schedule and the contractual or baseline finish date beyond which the expense of delay costs will start to kick in. Are we better off with a "looser" 25-day schedule and no schedule reserve or with a 21-day schedule and the schedule reserve? I know which I would want on my project investment!

Three-point estimating

No chapter on critical path scheduling would be complete without a few words about three-point estimating. Originally developed in 1958 by consultants at Booz Allen Hamilton, it was first used on the US Navy's Polaris missile program as part of what became known as the Program Evaluation & Review Technique or *PERT*. The term "PERT" is still used today, usually as a misnomer for the sort of CPM network logic diagram we have looked at previously. "Show me a PERT chart of that," a customer or executive is likely to say. But the chances are that what they are expecting to see has no relation to the original PERT chart.

CPM was developed in 1957 by consultants and engineers at the RAND Corporation and the DuPont Corporation for use on construction and plant maintenance projects. One year later those Booz Allen Hamilton consultants were trying to develop budgets and schedules for the Polaris program. From a planning point of view, what is the main difference between construction/plant maintenance projects and a program to develop a system that would launch a nuclear missile from a submerged submarine and hit a target thousands of miles away? The most obvious difference is that people have been doing construction and plant maintenance projects for a long time and have been able to collect a large database of historical data regarding durations and costs. Conversely, developing a submarine-launched missile system was brand new in 1958; there were no historical data that engineers could rely on for estimating purposes.

Under these circumstances, the consultants found it very difficult to pry estimates out of the engineers, who undoubtedly were smart enough to be reluctant about stating any numbers that could then be carved into concrete and used against them. This is often a problem with getting estimates, and leads to the generating of padded estimates. The padding then multiplies as it flows up the levels of the project, ultimately to infect and inflate both the duration and budget. Some of the negative effects of such padding are

- The sponsor decides not to do the project because it's likely to take longer and cost more than it will be worth.
- The customer looks at the proposal and decides to go with a different contractor.
- We undertake the project but, because work expands to fill both time and budget available (Parkinson's law), the project ends up being far less profitable than expected and perhaps even loses money.

The best way to avoid those negative effects is to base estimates on historical data. But even then, those data must be checked with the person who is going to be responsible for actually doing the work. It's pointless to insist that the performance data generated by very experienced engineers be matched by the novice you currently have on your team. Nor does yelling at your team member do any good. You have what you have—ask for a 50–50 estimate (equally likely to be over or under by a similar amount)— and if it's a lot more than previous actuals would indicate, ask why. It can be acceptable to ask your team member to "take a challenge" and try to get the work done within the lower estimates. But in such a case, be sure to add appropriate amounts to both schedule and cost reserve at the project level; you're likely to need them.

Three-point estimating is an attempt to formalize a technique that accounts for the probabilities of the estimates. It asks the estimators to provide not one but three estimates:

1. An optimistic, or minimum, estimate
2. A most likely estimate
3. A pessimistic, or maximum, estimate

One aspect of this method is immediately obvious: what do those terms really mean? Does pessimistic mean how long it will take and how much it will cost if things don't go quite as planned? Or does it mean taking into account the possibility that the activity might be interrupted by an evacuation for simultaneous earthquake, tsunami, and nuclear meltdown (which, sadly, has happened)? One estimator's optimistic estimate is a second person's most likely and a third person's pessimistic.

A second aspect is the fact that the optimistic estimate is almost always more constrained than the pessimistic one: there is a limit on how quickly one can make a hundred mile journey by car, but if there's a blizzard and visibility is zero and we get into an accident, we might literally never get there.

Three typical estimates for the work hours required to complete an activity might be

- Optimistic = 40
- Most Likely = 80
- Pessimistic = 180

We can plot these three estimates on a histogram as shown in Figure 5.12.

As mentioned earlier, it's hard to know exactly what is meant by the optimistic and pessimistic estimates: does it mean that the result could never be less than 40 or greater than 180? Some organizations and PMOs provide estimators with guidelines: "95% of the time" or "99% of the time." Of course, the estimates then exclude cases that are rare but which can have a great effect; are we ignoring the 0.5% of the time that the pessimistic should be 400 (perhaps reflecting a system that has to be reworked three times)? These are examples of what Nassim Nicholas Taleb termed "black swans" in his insightful 2007 book of that title. [2] In a project with 5,000 activities, a black swan that occurs 0.5% of the time would be expected to affect 25 activities and their schedule descendants.

But perhaps even more concerning is the most likely estimate; just how likely is it anyway? In the situation above, we presume that the estimator believes that no other result is more likely than 80. But is that 80 a result we can expect 25% of the time or 40% of the time or 75% of the time? It can make a considerable difference to our risk analysis.

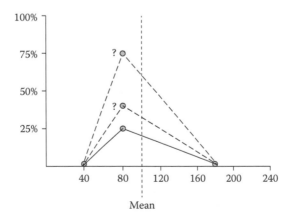

Figure 5.12 Optimistic, most likely, and pessimistic estimates plotted on a graph.

As I have noticed is often the case, the engineers on the Polaris program were happier giving three estimates instead of just one. Apparently the ability to include a "worst case" estimate makes people more comfortable. The Booz Allen Hamilton consultants then took the three estimates and used a formula weighted toward the most likely to compute what it would take. That formula became known as the PERT formula:

Estimate = (Optimistic + 4 * Most Likely + Pessimistic) ÷ 6

For our three estimates, this would lead to (40 + 4 * 80 + 180) ÷ 6 = 540 ÷ 6 = 90.

The PERT methodology then used the mean of the three estimates (i.e., 90 in this case) as the planned resource usage (or, in some cases, the duration) of the activity. Notice that this is 12.5% higher than the most likely estimate, a result of the pessimistic estimate being typically quite unconstrained. Thereafter, the scheduling theory divided into two camps: the probabilistics who supported three-point estimating and the deterministics who relied on one estimate. The probabilistics pointed out that no estimate is carved in stone and argued that if you only got one estimate, it would be the most likely one that would not even be the mean. The deterministics agreed that actual durations are probabilistic, but maintained that getting three estimates did little to improve things; asking for a "pessimistic" estimate merely formalized estimate padding. As one friend and longtime consultant once said to me: "If I tell you I'm going to have an activity finished on the 5th of the month if everything goes perfectly, but most likely you shouldn't expect it until the 12th and the way things go around here you probably shouldn't expect it till the 25th, what do I regard as my real drop dead date? When will I work late or come in on the weekend to get it finished?"

Another colleague voiced her concern more succinctly: "You're not going to believe one estimate but you'll believe three from that same person?"

The differing opinions remained in place until the computer came along...

Monte Carlo systems

Someone with a pocket calculator could quickly determine the means for hundreds or thousands of activities with three-point estimates and this may have had value in estimating work hours or cost. But it had severe limitations when dealing with durations and schedules. Schedules are driven by the critical path.

Within a schedule, there may be hundreds of merge points, where many predecessors merge into one successor, and burst points, where one activity precedes many successors. The longest path to each of these merge points determines the schedule for each of the activities following the burst.

But simply computing the mean of three estimates for each activity tells us little. If we have five different paths each of five activities all leading to a merge point, the path that turns out to have the most pessimistic results is likely to affect the merge and all its successors, and there's really no way to determine this without a computer.

Today there are many software packages that include what are called probability risk analysis modules or Monte Carlo simulations. You feed in the three estimates per activity, tell the software the distribution shapes, and the software will crank through the schedule of logical dependencies and duration estimates thousands of times to determine the likely length of the project. Most such packages will compute standard deviations and use Gaussian mathematics to print out a table of probabilities for completion dates: a 35% probability of finishing by July 15th, a 50% probability of finishing by July 27th, and 80% probability of finishing by August 13th, and so on. I have seen such tools used many times. They are wonderful for persuading customers that their expectations are unrealistic. After all, these predictions weren't just made up: they came out of a computer!

The purveyors of these Monte Carlo systems will, of course, do everything to persuade a potential buyer that the predictions are tremendously scientific. And no one really has a vested interest in pointing out the emperor's lack of clothing. But as with all computer systems, it's garbage in, gospel out. Here are some of the shortcomings that make the predictions of such systems arbitrary and unreliable:

1. The estimates are subjective. If they are generated by individual estimators, their personality differences affect the estimates differently: optimists estimate optimistically, pessimists estimate pessimistically. And formalizing guidelines for the estimators ("Make your pessimistic estimate what you should be able to achieve or surpass 98% of the time.") does little to help; an optimist's and a pessimist's 98% can be very different!

2. Even if the estimates are assembled from historical data, this subset of similar activities is chosen subjectively. Was the same activity manager in charge of all of them? Was the same team assigned to all of them? How much of the variation might be caused by the individual personnel? Was the work really identical? Who decides which previous activities are similar enough for inclusion as part of the basis of estimates (BOE) and which are not similar? There is a huge amount of room here for fudging the results.

3. A major shortcoming of most Monte Carlo simulation systems is the way that they handle schedule lag. Three estimates can be input for each activity, but the delay factor known as a lag is typically input as a single fixed amount. There is little awareness of this factor, but it can cause absurd situations. Let us take an example where we need

to dig a trench and then lay a pipe in the trench. We don't have to dig the trench the full quarter mile of its planned length before we start laying the pipe; after 4 days of trench digging, that equipment will have moved down the road and we will be able to start laying the pipe safely. Most schedulers would model this as what is called a start-to-start relationship with lag, that is, with a lag of 5 days between the start of digging the trench and the start of laying the pipe. Now let us suppose that our three estimates for the trench-digging activity are 7 days, 12 days, and 25 days. When the Monte Carlo software is running the simulation with the 12-day estimate, the lag will be assumed to be 5 days; and when it runs the simulation with the 7-day estimate, the lag will still be assumed to be 5 days; and when it assumes the 25-day estimate, the lag will be—that's right—5 days. Obviously, this is absurd. But in almost all Monte Carlo software, lags are treated as fixed amounts of time and not as volumes of work. And if there are a number of lags on the critical path, this can cause a significant distortion.

4. Perhaps the most distorting factor in the Monte Carlo simulation software is the question of what should really be the shape of the probabilistic distribution. Most of the software proudly offers a wide range of possible distributions, a menu of up to 30 different shapes that can be selected for each activity. Figure 5.13 shows what some of those distribution shape options might be. So let's see, we've got a medium-sized project, say 3,000 activities. How many project managers/schedulers/estimators do you think analyze each of their activities and try to figure out what the distribution shape ought to be? I suspect that if you guessed zero you would be close to correct. In fact, the vast majority of Monte Carlo risk simulation system users run it using a default distribution shape. By far the most common distribution shape used is the triangular distribution. The second most common distribution shape used is what is generally called the beta distribution, which is simply the output of a typical PERT formula estimation: (Optimistic + 4*[Most Likely] + Pessimistic) ÷ 6. These two distribution shapes are shown in Figure 5.14.

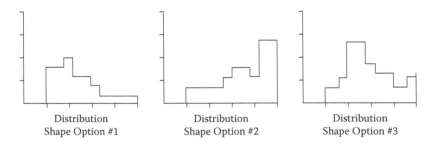

Distribution　　　　　　　Distribution　　　　　　　Distribution
Shape Option #1　　　　　Shape Option #2　　　　　Shape Option #3

Figure 5.13 Optional distributions for activity durations in Monte Carlo simulations.

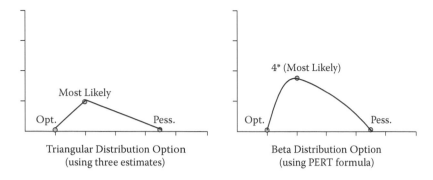

Triangular Distribution Option
(using three estimates)

Beta Distribution Option
(using PERT formula)

Figure 5.14 Two Gaussian distributions used as defaults in Monte Carlo simulations.

So how important is the distribution shape? And how much difference would it make if one first ran all the activity estimates using the triangular default and then ran them again using the beta default? Well, about three years ago I attended the annual conference of PMI's Scheduling Community of Practice. One presentation was by a very experienced gentleman who is quite prominent in the project management community. He recounted his experience on a particular project and talked extensively about the importance and value of using Monte Carlo simulation software to estimate project duration. During the question and answer session at the end of the presentation, I asked how he selected the distribution option he used for each activity. He replied that he ran it on one of the default distribution options. I asked which one and he said he wasn't sure, but that it really didn't make much difference. As this was a professional whom I respected, I was quite flabbergasted: the difference that can be expected for the 50% confidence level between using the triangular distribution default versus the beta distribution default ranges from 8% to 15%, and it increases as the confidence level rises to 60%, 80%, and so on. This may not seem like much, but remember that, through Parkinson's law, these estimates affect behavior. Imagine that the project is to implement a life-saving system within a medical center, one that will prevent fatal medication reconciliation errors. Further imagine that two people are dying every week that the project is delayed. If the project schedule had been run through a Monte Carlo risk simulation software package using the triangular distribution default, it might have suggested that we should use a deadline of 25 weeks. Now everyone works to that deadline, we deliver the system in 25 weeks, and everyone is happy. But what if we had run the software using the beta distribution and generated a deadline 8% shorter at the same level of confidence? Could we have finished the project in 23 weeks and saved four additional lives?

Which is the right distribution shape? I don't know the answer nor does anyone else. But to say that the distribution shape has little impact is dangerous nonsense. How many people a year die because the risk software is run on one default distribution instead of another? Of all possible reasons for dying, that seems like a singularly capricious one: the hellish interaction of blind trust in algorithms with veneration of deadlines.

If you are an executive with a contractor organization and you feel that your proposals set parameters that are too high and causes you to miss out on contracts, just mandate using the beta distribution default instead of the triangular distribution default; they're both arbitrary anyhow. (Or, alternatively, submit bids that de-emphasize deadlines and instead incorporate incentives for early delivery and penalties for late delivery.)

Let me be clear in summing up my feelings about three-point estimating and Monte Carlo risk simulation software. Does using these techniques and tools add something useful to our knowledge about, and ability to manage, the project? Yes, a little:

- Three-point estimating can perform as a sanity check when an individual estimate flies in the face of our intuition.
- It also can give us a sense of which activities may be most volatile in their duration.
- It's a really good tool for persuading a customer or sponsor who has been refusing schedule or cost reserve as part of the plan, or claiming that the project shouldn't take as long or cost as much as the project manager is saying it should. ("See, boss? It's not just me; even the computer says so!")

Justifying an estimated amount of schedule or cost reserve is certainly a valuable thing to be able to do. But believing that just because the Monte Carlo system says that we need precisely 18% schedule reserve in order to have a 75% chance of finishing the project on time is voodoo project management. There is nothing precise or particularly scientific about any of this. And it can require an enormous amount of effort, as well as the Monte Carlo simulation software package, which is not cheap.

Is collecting all the estimates for thousands of activities, selecting what seems like the right distribution shape for most, and then running the software through 10,000 simulations likely to produce a reliable schedule reserve estimate? I'm honestly not sure that it's significantly more reliable than a process that allots 10% reserve for a low-risk project, 20% for a medium-risk project, 30% for a high-risk project, and 40% for a very high-risk project. But then again, I'm not a computer.

Summary points

1. Time taken on a project almost always reduces its business value. If the value/cost of that time earlier or later than a target completion date is estimated, that metric can be driven down to the level of the individual critical path activity. If an activity has two weeks of critical path drag and each week earlier that the project is completed would be worth $25,000, the drag cost of that activity is $50,000. This can be used to justify the resources needed to reduce an activity's drag.

2. The true cost of an activity is the sum of its resource costs plus its drag cost. Because only critical path activities have drag and drag cost, the true cost of doing critical path work is usually much greater than doing work off the critical path. And the true cost of an activity can often be reduced by increasing the budget to reduce the drag cost and hence the true cost.

3. On a contractual project, drag computation combined with knowledge of the value/cost of time to the customer can provide the contractor with an opportunity to amend the contract to increase value to the customer and to the contractor through incentives for earlier delivery.

4. Activities may or may not get shorter (or sometimes longer!) if the resourcing levels are increased. This is called an activity's resource elasticity, and it is dependent on the specific work context of the activity. The doubled resource estimated duration is a secondary duration estimate that can help point the project manager to specific activities whose drag and drag cost would be most reduced by additional resources.

5. In any project-driven organization, where the majority of revenues are generated through projects, personnel in certain departments who can have a great impact on project performance are often untrained in crucial project management knowledge such as critical path analysis and drag cost. This is a very costly problem in many organizations and needs to be addressed. Almost all members of such organizations need to be trained in project management.

6. Three-point estimating techniques and Monte Carlo systems can add value, but they are not a panacea for more accurate estimating. Users of such systems need to understand fully the theory behind them and what their limitations are.

Endnotes

1. Sometimes such compression may not even involve additional resources and cost, but simply the compression technique called *fast tracking*, changing the network logic so that a successor activity can start after only a portion

of its predecessor is done, either by decomposing the predecessor into its component parts or by using a start-to-start (SS) relationship (which as I mentioned earlier, is beyond the scope of this book, but is covered thoroughly in my previous book, *Total Project Control*). We can accomplish the same thing by decomposing the predecessor activity and placing a finish-to-start relationship between the first part of the predecessor and the successor. However, it is important to recognize that fast tracking is no panacea: it both makes the network more complicated and increases risk by requiring new activities to start on the basis of less information.

2. Nassim Nicholas Taleb, *The Black Swan: The Impact of the Highly Improbable*. New York: Random House, 2007.

chapter six

Combining project investment tools

"How do we use these tools to maximize a project's business value?"

Using the techniques we have been describing, although they are of significant and demonstrable value, requires changing the way an organization thinks about projects. Having leadership and support at the senior management level is crucial. Why is this so challenging?

- "We've been doing just fine for years!" Actually, it's likely that the sense of "doing just fine" is an illusion due to the lack of investment metrics that would show how much money and value has leaked away. If things such as the cost of time are left unquantified, it's easy to think that a project that delivered the intended product by an arbitrary deadline was "successful," even if completing it in half the time was feasible and would have added greatly to the business value.
- These new metrics will often require guidance about project and time value from project sponsors and business analysts. Many leaders have been successful because they have used their judgment, not necessarily because they have been data driven. This kind of thinking, making judgments based on observations, experience, and industry knowledge, without analysis and metrics, will not be optimal for projects. Daniel Kahneman describes this in his 2011 book, *Thinking, Fast and Slow*, as "fast" thinking, which is absolutely appropriate in certain management decisions, but not the type we are discussing here. Kahneman points out that "slow" thinking often requires calculation, an uncomfortable effort that most human beings try to avoid whenever possible. He uses behavioral studies to show how people will opt almost every time for the answer based on judgment rather than data when faced with an issue that requires adopting an analytical approach.[1] And that judgment is often based only on the superficial.
- Project management as a discipline is not well known. Many leaders in a project-driven organization don't fully understand the workings of the underlying concepts such as work breakdown structures, critical path and resource schedules, and earned value tracking,

therefore embarking on an effort to understand slightly more complex enhancements to these techniques is quite intimidating.

- Although there is also concern, being required to make estimates and to embrace a system with trackable metrics might lead to improved organizational results, it will highlight bad forecasts, bad decisions, or costly and inefficient results that may embarrass those managers who made them.

In many organizations, only a mandate from the executive level that requires (or empowers the PMO to require) specific data at project initiation (i.e., during development of the project charter, business case, or contract) may suffice to impel the organization to start making the effort to analyze and manage the projects as investments. And for any large project investment, these value estimates should not only be tracked but also be checked both at the end of the project and again at some appropriate time after the project's product has been launched.

Yet implementing these metrics and techniques will lead to significant improvement in project results, as measured in what should be the most important metric: expected value generated above invested cost. All too often, programs, projects, and even portions of projects are undertaken that a little upfront analysis would show are either not worth doing or could be done much better.

In the rest of this chapter, through the use of a sample project, we show how to use investment value along with drag and drag cost both to optimize a schedule during upfront planning and to recover a schedule that has slipped. The format is based on a step-by-step process through the different data items and techniques that we need to use as we plan and optimize the schedule. We use as an example the training project shown in Figure 6.1. We refer to the durations as though they are in days, but if they were in hours or weeks, the principle would still be the same.

The sample network for a training project is deliberately chosen to be as generic as possible. Readers can assume that the training is for whatever purpose seems most meaningful to them. Some possible training projects are

a. For a medical team providing immunizations against a recent disease outbreak
b. For users of a new corporate application
c. For a military unit in the use and maintenance of a new defense system

As the immunization project introduces some interesting factors, I use it as the primary model. But the main methods referred to below are applicable in almost any type of project, in any field.

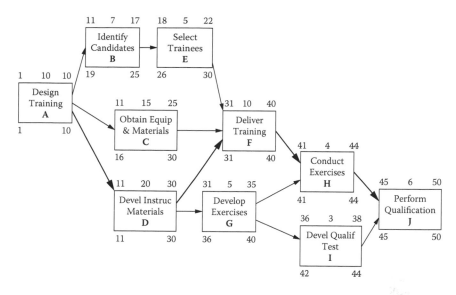

Figure 6.1 Simple network diagram for a training project.

Step 1: Determine expected monetary value of the project

The individual members of the project team do not necessarily need to know the project's EMV. But such information can certainly be valuable to the project manager, and should be absolutely crucial to the sponsor/customer and to the sponsor/customer's organization. There are projects that are undertaken where, if everything went according to plan, the expected value generated would be less than the budget. This is obviously an unsatisfactory situation, but it is one that will continue unless and until:

- The sponsor is required to estimate the expected value and itemize the specific value generators and their contributions. (For example, if the project is an enabler project within a larger program, what are the values it will enable?)
- There is a formal review scheduled for an appropriate time after project completion to see whether the project's expected value was in fact generated or if the sponsor's estimates were wildly optimistic.

The first step in avoiding future bad investments is to measure the value of investments in the past.

The following case is constructed to illustrate the concepts and enable calculations. To flesh out the details of our sample project:

1. The immunization program is to protect against a recent outbreak of a disease in a secluded area of a Third World country. The affected area currently has a population of 10,000 people. If nothing is done, the lethality of the disease suggests that 20% of that population, or 2,000 people, will die. Twenty people have already died, and currently one more person is dying every day.
2. The assumption is that if the current trends continue, this number of fatalities will double every 10 days: to two per day from Day 11 to Day 20, four per day between Day 21 and Day 30, and so on. Between Day 71 and Day 80, the number of daily deaths will taper off as only those with a natural immunity remain. The average death rate in that last 10-day period would be about 71 per day. Then the mortality should cease.

Figure 6.2 displays a damage control time chart if no medical response project is implemented. This chart measures damage in lives lost, and is a vital document for emergency response projects. However, a similar document measuring the damage of delay in revenues or savings lost over time would be the relevant measure for any project.

The plan for the training project (as shown in Figure 6.1) to support the immunization program details the following assumptions:

1. The team can start the training project immediately and begin immunizations as soon as training is completed and the vaccine is ready. The vaccine that the immunization team will use is expected to be ready at the end of Day 40. This means that if the training project takes longer than 40 days, each of those days will be drag, delaying implementation of the immunization program.
2. Daily deaths are predicted to be cut in half 10 days after vaccinations begin, and to cease 20 days after vaccinations begin.
3. The government has mandated that, for planning and budgeting purposes, a figure of $5 million should be used in computing the cost of each death and value of each life saved.

	Deaths so Far	Days 1–10	Days 11–20	Days 21–30	Days 31–40	Days 41–50	Days 51–60	Days 61–70	Days 71–80
Daily Deaths if No Response	20	1 per day	2 per day	4 per day	8 per day	16 per day	32 per day	64 per day	71 per day
Cumulative Deaths if No Response	20	30	50	90	170	330	650	1290	2000

Figure 6.2 Initial damage control time chart with no immunization program.

Based on this information, we conclude that if we complete the project at the end of 50 days and begin the immunizations immediately after, 330 people will already have died (20 + 10 + 20 + 40 + 80 + 160) by the time we start immunizations. Thirty-two people will die each day of the first 10 days of immunizations between Day 51 and Day 60, a total of 320 more deaths. But the death rate will be halved in the following 10 days. An average of just 16 people will die each day from Day 61 to Day 70, for an additional 160 deaths. And with the vaccines taking full effect after Day 70, there should be no more deaths. The total number who will die before immunization takes 100% effect will be 330 + 320 + 160 = 810, 1,190 less than if we simply wait for the disease to run its course. At $5 million per life saved, the value of this project if completed in 50 days so that immunizations can start on Day 51 will be 1,190 * $5 million = $5.95 billion. That is the expected monetary value of the entire immunization program of which the training project is a part. The new damage control time chart with the impact of the immunization program as initially planned is shown highlighted in Figure 6.3.

This training is an enabler project: training itself will save no lives, but without the medical responders training project, many more will die. Our analysis indicates that if we don't do the planned training, it will take considerably longer to disseminate the vaccines. Instead of 16 deaths per day between Day 61 and Day 70, there will be 24 deaths per day, an increase of 50% or 80 more deaths. In addition, instead of the death rate going to zero between Days 71 and 80, it will still be 20 deaths per day, or another 200 deaths. Thus the training project is adding value equal to the saving of 280 human lives to the immunization program. The value-added in dollars of the training project is 280 * $5 million, or $1.4 billion.

	Deaths so Far	Days 1–10	Days 11–20	Days 21–30	Days 31–40	Days 41–50	Days 51–60	Days 61–70	Days 71–80
Daily Deaths if No Response	20	1 per day	2 per day	4 per day	8 per day	16 per day	32 per day	64 per day	71 per day
Cumulative Deaths if No Response	20	30	50	90	170	330	650	1290	2000
Daily Deaths w/ Current Plan	20	1 per day	2 per day	4 per day	8 per day	16 per day	32 per day	**16 per day**	**0 per day**
Cumulative Deaths w/Current Plan	20	30	50	90	170	330	650	**810**	**810**
Lives Saved w/ Current Plan	0	0	0	0	0	0	0	**480**	**1190**
$ Value of Lives Saved at $5M Each								**$2.4B**	**$5.95B**

Figure 6.3 Damage control time chart with initial plan for immunization program.

Step 2: Develop a value breakdown structure (VBS)

A valuable tool in assessing a project's expected value is a *value break-down structure* (VBS). This is a document that I introduced and explored in depth in my 1999 book, *Total Project Control*. To explain briefly, a VBS lays out the value-added of each component or work package within the project, that is, how much each contributes to the total expected monetary value. However, a VBS does not have to be limited to the project level. It is of even greater value at the program level where the contributions of each project within the program can be defined, so that we can see how the work of an enabler project is adding value to the program that would not exist without it.

Although a VBS is hierarchical and looks very much like a work breakdown structure (indeed, at the project level the format of the VBS can be identical with that of the WBS), there is one crucial difference: whereas data such as cost that are entered into the lower levels of a WBS can then be summed up each branch all the way to the top, that is not the case with a VBS. Value is not additive in the way that cost is. For example, let us imagine that we are building a six-passenger airplane. The budget is $100,000 and the EMV is $150,000. We might decompose the WBS into specific components: fuselage, propulsion system, instrumentation and avionics, wings, landing gear, and tail section. If the cost of resources for building and attaching each wing is $10,000, and we have completed the entire plane except for the left wing, we might expect to have spent $90,000, all except the budget for the left wing. But what is the value of the plane? Until the left wing is added the value of the plane is close to zero (well, perhaps a few dollars as scrap metal). Even if someone buys the plane from us without the left wing, it is only because they intend to add a left wing. The value-added of the left wing is equal to the entire value of the project because it is a mandatory activity; we have not done the project and cannot obtain its value until we have performed all mandatory activities.

Within any project, we will have mandatory work and activities and optional work and activities:

- *Mandatory* work has value equal to that of the entire project, and must be performed because (1) the project makes no sense without it, (2) the project has no value without it, or (3) it is required by law or by contract.
- *Optional* work does not have to be performed, but the project's expected value may be greatly diminished if it's not; that is, the work has a lot of value-added.

It is very important that the project manager know what work is mandatory and what work is optional. And, interestingly, it is the optional work that provides opportunity: mandatory work has to be done no matter what, but optional work cries out for cost–benefit analysis.

Let us look at two possible optional activities in our airplane project:

1. The instrumentation includes an autopilot.
2. It has been designed to include drop-down tables on all the seat-backs in the cabin.

An autopilot is a very valuable option to have on an airplane. If we're assuming we will be able to sell the completed airplane for $150,000 with a working autopilot, not including one might reduce that value by as much as $40,000. Thus the autopilot, although optional, has a large value-added. If its inclusion is budgeted for $8,000 and installation is on the critical path with drag of 5 days at an estimated drag cost of $4,000 per day, the true cost to the project of the autopilot is $28,000 ($8,000 + 5 * $4,000). This makes its net value-added a healthy $12,000 ($40,000 – $28,000). Unless something changes, such as the cost to install the autopilot increases dramatically or an installation delay increases the drag to 8 days or more and the drag cost to $32,000 or more, it makes sense to include the autopilot system. But notice that the potential increase in drag cost is a particularly insidious peril. If the path of autopilot activities slips, we could include in our project a system that costs us (in terms of its true cost) more than the value it adds.

The situation with the drop-down seatback tables seems simpler. Such seats will cost us $3,000 more than similar seats without the tables, and we estimate they will justify an extra $5,000 on the price. All is therefore fine, provided there is no danger of the seats winding up on our critical path. We check our schedule and all seat-related activities have at least 9 days of float, so we think we're okay. And then we get word from our supplier that manufacture of the seats with drop-down tables is being delayed by 10 days. Now our critical path will change and instead of the seat instal-lation having 9 days of float, it will have one day of drag with drag cost of $4,000. Now the true cost of our drop-down seats as opposed to those without tables is $3,000 + $4,000 of drag cost. If the value-added is only $5,000, the net value-added is –$2,000. We need to change our order to the seats without the tables.

Hundreds of projects are being performed every day that include optional work that is costing more than it's worth. The reason is the perni-cious interaction of six omissions:

1. Projects are not being analyzed as investments.
2. Investment-oriented metrics are not being used to track project performance.
3. Project managers are not trained in managing a project as an investment.

4. Value breakdown structures are not being developed and used.
5. The value-added of optional activities is not being estimated.
6. The drag cost and true cost of activities are not being computed, nor tracked as a project is performed.

Until all of these omissions are corrected, projects will continue to leak investment value through activities whose true cost is greater than their value-added.

During the planning and scheduling phase, we should estimate the value of components or work packages and make sure that any optional work always has a value-added that exceeds its true cost. During project execution, our project management software should allow input of both value-added and cost-of-time data so that it can automatically signal us if any activity's net value-added goes negative, or even drops below a threshold level. Lacking such software, we need to generate periodic reports on any activity whose net value-added is close to breakeven, and senior management and sponsor/customers should press to see such reports.

The net value-added of activities that start out on the critical path, such as the autopilot system, can become negative due to either cost or schedule overruns, or even due to acceleration on parallel paths that increase their float and may thereby increase the drag of the critical path activities. This is certainly something that a project manager should watch. But an even greater problem can arise any time that the critical path changes. Work that has a low value-added but used to be off the critical path and was therefore included now potentially has drag and drag cost, perhaps very expensive drag cost. Its net value-added can suddenly become dramatically negative. Our airplane example uses a delay cost of just $4,000 per day, but many projects have costs in the hundreds of thousands of dollars per day. On such projects, trivial optional activities can wind up costing millions of dollars if they migrate to the critical path.

Estimating the value-added of the training project to the immunization program

The glossary of the fifth edition of the *PMBOK Guide* (2013) defines *program* as "A group of related projects, subprograms and program activities managed in a coordinated way to obtain benefits not available from managing them individually."[2] Just as a project is scheduled by sequencing the work activities, a program is scheduled by sequencing the projects, subprograms, and program activities. But whereas sequencing the work activities in a project consists of tracing the physical logic to create the final product or service, sequencing the projects in a program is more discretionary, less driven by the laws of physics and more by the financial considerations of maximizing the program's value.

Within the program, each project is selected for the value it's adding. In the case of our immunization program, the training is an enabler project. Training itself will save no lives, but without good training for the medical responders, many more will die. Our cost–benefit analysis indicates that if we don't do the planned training, it will take considerably longer to disseminate the vaccines. Instead of 16 deaths per day between Day 61 and Day 70, there will be 24 deaths per day, an increase of 50% or 80 more deaths. In addition, instead of the death rate going to zero between Days 71 and 80, it will still be 20 deaths per day, or another 200 deaths. Thus the training project is adding value equal to the saving of 280 human lives to the immunization program. This is displayed in Figure 6.4.

Saving 280 additional lives (at an estimated dollar value of $5 million each, or $1.4 billion) is obviously very valuable. However, we are missing a significant implication of including the training project: the time it will add to the program and the deaths that added time may cause, otherwise known as the project's drag cost!

Just as the drag of activities on a project's critical path delays the creation of the project's product or service, the generation of a program's value can be delayed (and reduced) by a project on the program's critical path. You may recall we said earlier that the vaccines would be all ready to go at the beginning of Day 41. However, immunizing the population cannot start until the beginning of Day 51 because of the training project. That means that the training project has 10 days of drag on the program's critical path. How many deaths will that additional 10 days of delay cause? In other words, what is the drag cost of the training project?

Figure 6.5 shows the comparative damage control time chart if the immunizations could start on Day 41 instead of Day 51. The difference in the number of deaths represents the drag cost of the training project on the immunization program's critical path. In this case, starting

	Deaths so Far	Days 1–10	Days 11–20	Days 21–30	Days 31–40	Days 41–50	Days 51–60	Days 61–70	Days 71–80
Daily Deaths w/ Current Plan	20	1 per day	2 per day	4 per day	8 per day	16 per day	32 per day	16 per day	0 per day
Cumulative Deaths w/Current Plan	20	30	50	90	170	330	650	810	810
Daily Deaths w/o Training Project	20	1 per day	2 per day	4 per day	8 per day	16 per day	32 per day	24 per day	20 per day
Cum. Deaths w/o Training Project	20	30	50	90	170	330	650	890	1090
Addit. Deaths w/o Training Project	0	0	0	0	0	0	0	80	280

Figure 6.4 Damage control time chart of immunization program without the value-added of the training project.

	Deaths so Far	Days 1–10	Days 11–20	Days 21–30	Days 31–40	Days 41–50	Days 51–60	Days 61–70	Days 71–80
Daily Deaths w/ Day 51 Delivery Start	20	1 per day	2 per day	4 per day	8 per day	16 per day	32 per day	16 per day	0 per day
Cum. Deaths w/ Day 51 Delivery Start	20	30	50	90	170	330	650	810	810
Daily Deaths w/ Day 41 Delivery Start	20	1 per day	2 per day	4 per day	8 per day	16 per day	8 per day	0 per day	0 per day
Cum. Deaths w/ Day 41 Delivery Start	20	30	50	90	170	330	410	410	410
Lives Saved w/o Training Project's Drag	0	0	0	0	0	0	240	400	400

Figure 6.5 Damage control time chart with immunizations starting 10 days earlier (without the training project's drag).

the immunizations when the mortality rate is at a level of 16 per day means that the level never reaches 32 per day. Instead, as before, after 10 days the mortality rate is cut to 50%, meaning 80 deaths between Day 51 and Day 70 rather than the 480 deaths that would occur if we don't start immunizations until Day 51.

So now we can see that although the training project's scope is projected to save 280 lives, it's presence on the program's critical path currently gives it drag of 10 days at a drag cost of 400 lives. All else being equal, we would save 120 more lives by eliminating the training project than by performing it on the current schedule. This gives the training project a net value-added of minus 120 lives in the current program plan, or –$600 million (not including the budget for resources). Should we do the training or not?

Estimating the value-added of activities in the training project

The drag of a project in a program is driven by the critical path of activities in the project. In order to reduce the drag and drag cost of the training project, we should analyze the drags and drag costs of the activities on the project critical path. We might be able to do this by adding resources or by doing more work in parallel (fast tracking). But for this example, let us repeat at the project level the sort of value-added analysis we just did at the program level.

First, we determine that the mandatory and optional activities for our training project are as shown in Figure 6.6. As we can see, Activities A through F are about development and delivery of the training and are

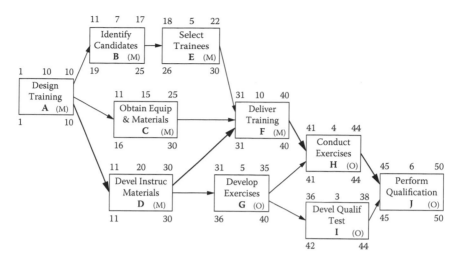

Figure 6.6 Mandatory and optional activities in the training project.

mandatory (although if we decomposed each to a finer level of detail, we might find work, such as obtaining specific training materials, that would be optional). The last four activities, involving exercises, testing, and qualifying of the trained personnel, are optional. We could just deliver the training without practical exercises or testing and hope that information has sunk in.

Although Activities G (Develop Exercises) and H (Conduct Exercises) are optional, each has a high value-added, practicing the application of the lessons learned from the training in a realistic manner. Without these exercises, for many members of the immunization team their first few days will essentially be "on-the-job training." Many may not initially be proficient, and the result could be more deaths during the early stages of the immunization program. Let's assume that, without the experience provided by the simulation exercises, instead of deaths after the first 10 days of the program being cut by 50%, they will only be cut by 25%. Thereafter, the team will be experienced enough that there will be no further change; deaths should cease 20 days after immunizations begin.

From Day 51 to Day 60, 32 people per day were expected to die without the immunization program. That number was to be reduced by half from Day 61 to Day 70 by the immunizations as scheduled, to 16 people per day. But if the exercises are not included in the training, that number will only be cut by 25%, to 24 people per day, from Day 61 to Day 70. The result will be that 80 more people will die, raising the total death toll from 810 to 890. At $5 million per life, the immunization program will save only 1,110 lives at a value of $5.55 billion instead of 1,190 lives at a value of $5.95 billion. Therefore the two activities to develop and conduct the exercises (G and H) have a value-added of 80 lives, or $400 million, well worth including in the training project, most people would instinctively say.

Activities I (Develop Qualif Test) and J (Perform Qualification) are certainly not mandatory, yet they do add some value. Without these activities, some personnel who are likely to make errors will be conducting immunizations. As with Activities G and H, we should analyze what the impact of those errors might be. Let's presume that we do so and discover that the evidence from similar immunization programs suggests that a small percentage of the vaccines may be administered improperly, resulting in an average of one additional death per day from Day 61 to Day 70, or a total of 10 additional deaths. But who would not want to save those 10 additional lives? This analysis enables the decisions, which are always difficult due to competing priorities and limited budgets, to be driven by data.

Step 3: Determine value/cost of time on the project

As I mentioned earlier, insisting on the need for the project sponsor/customer to estimate the value/cost of time on the project is one of the most important things that any executive can do to improve his organization's project management effectiveness. But if the sponsor/customer (or in this case, the program manager) does not provide this information to the project manager, then the project manager is well advised, for the benefit of the project, to do such an analysis anyway.

In the specific case of the immunization program, the value of time in lives lost or saved varies according to the completion date. The scope of the entire training project is currently saving 280 lives, but its drag of 10 days within the program is currently costing 400 lives. Without further analysis, it might be seen as an easy call to jettison the training program and save approximately 120 additional lives.

But on programs and projects, both the gods and the devils are in the details. The project manager of a key project (like an enabler) on any program needs to do cost–benefit analysis of the project's scope. On this training project, the project manager determines that the scope of developing and conducting the practical exercises (Activities G and H) has a value-added of 80 lives saved. The scope of developing the qualification test and administering it (Activities I and J) has a value-added of 10 lives saved. Should these optional activities be included in the project scope? We cannot know the answer without knowing their drag costs.

Step 4: Computing drag cost of optional activities

From the program, we know that reducing the duration of the training project by 10 days will save 400 lives. And from the project's critical path schedule, we know how much time each activity is adding to the project duration, that is, its drag. Knowing the drag cost at the program level now allows us to compute the drag cost at the project and activity level, in human lives lost. This metric should always be computed on any project intended to save human lives in homeland security, law enforcement, emergency response, potable water wells, hospital construction, pharmaceutical and medical device development, and of course immunization programs.

Figure 6.7 shows the immunization project schedule with the total float or critical path drag of each activity computed. Activities G and I are off the critical path with total floats of 5 days and 6 days, respectively. Activity H is on the critical path with a drag of 4 days because even though it is parallel to Activity I, which has a total float of 6, Activity H's drag is limited by its duration of 4 days. Activity J has nothing in parallel and so its drag is equal to its duration of 6 days.

What are the drag costs of Activity H and Activity J in the current schedule? Let us take Activity J first, both because it has more drag and because it has less value-added. If Activities I and J were both eliminated, their combined drag of 6 days (all in Activity J) would be removed from the training project and it would finish at the end of Day 44. This would eliminate 6 days of the training project's drag on the immunization program and allow the immunizations to start on Day 45. As shown in Figure 6.8, the death rate would therefore be reduced starting on Day 55.

Instead of the full 10 days, from Day 51 to Day 60, of 32 people dying every day, that would now be the death rate for only 4 of those days, from the beginning of Day 51 through the end of Day 54. From Day 55 (10 days after the immunizations began) to the end of Day 64, the death rate

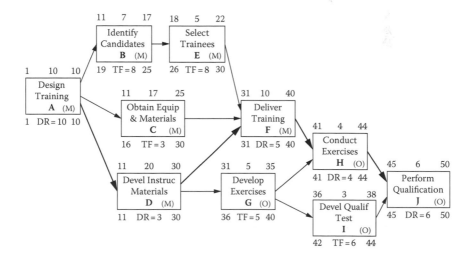

Figure 6.7 Training project showing optional activities and drag totals.

	Deaths so Far	Days 1–10	Days 11–20	Days 21–30	Days 31–40	Days 41–50	Days 51–54	Days 55–60	Days 61–64	Days 65–70
Daily Deaths w/ Day 51 Delivery Start	20	1 per day	2 per day	4 per day	8 per day	16 per day	32 per day	32 per day	16 per day	16 per day
Cum. Deaths w/ Day 51 Delivery Start	20	30	50	90	170	330	458	650	714	810
Daily Deaths w/ Day 45 Delivery Start (NO I or J)	20	1 per day	2 per day	4 per day	8 per day	16 per day	32 per day	**16 per day**	16 per day	**0 per day**
Cum. Deaths w/ Day 45 Delivery Start (NO I or J)	20	30	50	90	170	330	458	**554**	618	**618**
Lives Saved w/o Drag of Activities I and J	0	0	0	0	0	0	0	96	96	192

Figure 6.8 Damage control time chart without the drag of Activities I and J.

would be reduced by 50% to 16 per day. This would result in the saving of 96 (i.e., 6 times 16) lives. In addition, 20 days after the immunizations start, or beginning with Day 65, deaths from the disease would cease. This would result in the prevention of 16 deaths per day from Day 65 to Day 70, an additional 96 lives. The drag cost of Activity J is 192 human lives. (It seems blasphemous even to do the calculation, but at the mandated value of $5 million per life, that represents a monetary saving of $960 million.)

For Activity H, its drag is only 4 days. If its duration were not in the schedule, the project would finish after Day 46 and we could begin immunizations at the start of Day 47. The death rate would therefore be affected by the immunizations starting on Day 57. We can do the calculations in the same way as we did for Activity J's 6 days of drag. If we do so, we see that from Day 57 (10 days after the immunizations would begin) to the end of Day 60 the death rate will be reduced to save 64 (i.e., 4 times 16) lives. In addition, 20 days after the immunizations start, or beginning at the end of Day 66, deaths from the disease will cease. This will result in the additional prevention of 16 deaths per day from Day 67 to Day 70, or another 64 lives saved. The drag cost of Activity H is 128 human lives.

Step 5: Calculating net value-added of optional activities

In these four optional activities, we have two pairs: in each case, one is off the critical path and has float and one is on the critical path and has drag and drag cost. Yet common sense tells us that Activities G and H must be considered together because the whole purpose of Activity G

(Develop Exercises) is to perform Activity H (Conduct Exercises), and we cannot perform Activity H unless we have first performed Activity G (or, if you will, that Activity G is an enabler of Activity H). Similarly, Activities I (Develop Qualif Test) and J (Perform Qualification) must be paired: the only purpose in developing the qualification test is as an enabler of Activity J, perform the qualification process.

Although Activities G and I have no drag costs we would need to include their resource costs in analyzing the true costs for implementing the exercises (Activities G and H) and implementing the qualification process (Activities I and J). However, in our training project for the immunization program, the value of the lives we are saving and in which we are measuring drag cost will obviously far outweigh the cost of resources.

Back in Step 2, we analyzed that the value-added of doing the exercises would be the saving of 80 additional lives and of doing the qualification testing would be 10 lives. But because some of this work is on the critical path of a project that is on the critical path of a program, it is adding time to the program duration that reduces the value generated by the program.

As we saw in Step 4, the cost of Activity H's 4 days of drag is 128 lives. That gives us a net value-added (or, if you will, a net value-subtracted) for Activities G and H of minus 48 human lives: it will cost 48 additional human lives to perform the testing. We are better off starting the immunizations earlier with personnel for whom better training would've saved 80 lives but cost 128 lives because of the added time.

Similarly, implementing the qualification process in Activities I and J will have a drag of 6 days and a drag cost of 192 lives. But in Step 2, we saw that the value-added of the qualification process was only 10 lives saved. Thus Activities I and J also have a negative net value of 182 lives.

Unless we can figure out a way to reduce the drags of Activities I and J, we should jettison the qualification process. Because of time on the critical path, it is costing more than it is worth, and hurting more people than it is saving.

Uncertainty principle in schedule optimization

One thing about project scheduling that is crucial to keep in mind: when you change one item in one part of the schedule, it can often have an impact on activities elsewhere. Change the duration of any activity and it can not only change the critical path; it is also possible that it will change float and drag calculations on other activities. In addition, the cost of time can vary depending on precisely which days or weeks are being affected by a schedule change. A schedule compression that pulls the delivery date of our *Softy the Snowboy* product back from December 2nd to November 25th allows us to access the Black Friday retail market; a similar compression that pulls us back by another week to November 18th would not be worth as much.

In our immunization program, once we eliminate Activities I and J and the resulting drag of 6 days, we have optimized the schedule to the tune of 182 additional lives saved. However, we have also changed the implications of eliminating the drag of the testing activities, Activities G and H. Activity H still has drag of 4 days (elimination of other activities can sometimes change the drag totals, although not in this particular case). But once we have eliminated Activities I and J, the key dates for immunization delivery all changed and with them the implications for saving time and lives through even earlier delivery.

Whereas previously (with Activities I and J still in the schedule), eliminating the 4 days of drag on Activity H would only have allowed immunization to start on Day 47 (instead of Day 51), once we have jettisoned Activities I and J, further compressing the schedule by eliminating Activity H pulls the start of immunization from Day 45 to Day 41. The damage control time chart in Figure 6.9 shows the difference in lives saved between starting immunization on Day 45 (without Activities I and J) and starting it on Day 41 (also eliminating Activities G and H).

Whereas the drag cost associated with doing Activities G and H was previously 128 deaths, the new schedule without the qualification process has increased the cost of 4 days of drag by 80, to 208 deaths. With the value of the scope of the exercises still 80 lives saved, the program or project team can now save 128 human lives by eliminating the tests. But only if they are aware that time on the critical path can have a cost much greater than the value of the work it produces.

	Deaths so Far	Days 1–10	Days 11–20	Days 21–30	Days 31–40	Days 41–50	Days 51–54	Days 55–60	Days 61–64	Days 65–70
Daily Deaths w/ Day 45 Delivery Start (NO I or J)	20	1 per day	2 per day	4 per day	8 per day	16 per day	32 per day	16 per day	16 per day	0 per day
Cum. Deaths w/ Day 45 Delivery Start (NO I or J)	20	30	50	90	170	330	458	554	618	618
Daily Deaths w/ Day 41 Delivery Start (NO G, H, I or J)	20	1 per day	2 per day	4 per day	8 per day	16 per day	**8 per day**	**8 per day**	**0 per day**	0 per day
Cum. Deaths w/ Day 45 Delivery Start (NO G, H, I or J)	20	30	50	90	170	330	**362**	**410**	**410**	410
Lives Saved w/o Drag of Both Activities H and J	0	0	0	0	0	0	**96**	**144**	**208**	208

Figure 6.9 Damage control time chart without the drag of Activities G, H, I, and J.

Final word on the schedule optimization process

The sample project was chosen specifically as a dramatic example of the way that inadequate analysis or bad decision making can and does cost humanity. But it also costs governments and shareholders and executives and managers and team members every day, in public welfare and in business.

If a project is indeed an investment, we must ensure that the work activities that comprise the elements which generate the value of that investment are worth their inclusion. Any activities that represent leaks of value can turn what should be very profitable projects into losses. Every day, projects are being delivered that include components and work that add less value than they cost. Sometimes the initial project plan was at fault: cost–benefit analysis should have shown, as with Activities I and J, the negative effects of including certain work items. But even more often, the problem does not arise until later, when the critical path changes.

Any time that the critical path changes during project performance, it requires renewed analysis of the value-added and the true cost of any activity that is part of the new critical path. And optional activities that made perfect sense to include when they were not on the critical path may no longer justify their inclusion.

In summarizing, for any optional activity on the critical path, either during upfront planning or when new activities migrate to the critical path, the customer/sponsor or senior management must ask for reports on

1. The estimated value-added of all such activities
2. The critical path drag and drag cost of those activities
3. The net value-added of any optional activity on the critical path, starting with those closest to zero

Those with negative value-added should be jettisoned if no way can be found to reduce their true cost. Every program and project manager should also take a proactive stance on this. Instead of waiting for the sponsor/customer to notice that a low-value component or work activity is on the critical path and is delaying project completion, the project manager should include true cost information for all critical path activities that appear to be optional, both when submitting the initial plan and at regular progress meetings. The project manager should also prompt the sponsor/customer to estimate the value-added of such components or activities and thereafter generate net value-added reports on them.

Using critical path drag to recover a schedule

Occasionally over the years some project managers have expressed doubts about the need to optimize a schedule in the planning phase. Their words vary, but the essence of their point is: "I'm given a scope of work, a deadline, and a budget, and my job is just to do what I'm told within the time and cost parameters I'm given. All that ROI and investment stuff is well and good, but that's for senior management or the customer to worry about. I've got enough on my hands just trying to deliver the technical aspects of the project."

I have sympathy for this position: why should the project manager and team worry about the investment aspects of the project if senior management, the sponsor, or the customer doesn't? The organization needs to shift to requiring careful analysis of such data and implanting them in the metrics in a way that will guide the project during planning and execution. As I have already discussed, if senior management wants to optimize the investment value of the project, it needs to mandate and disseminate that expectation throughout the organization through standardized processes and metrics.

Of course, there is no doubt that using many of these techniques can help the project manager even in cases where senior management neither mandates nor understands their uses. Using critical path drag to compress the working schedule and thus increase schedule reserve can save a project manager many sleepless nights when her schedule starts to slip a bit; she knows she still has sufficient schedule reserve in her back pocket.

Additionally, a project manager's informal estimate of the value/cost of time can be used in conjunction with drag and drag cost to justify additional resources for the project: "On the basis of marketing's revenue predictions for this new smartphone we're developing, I did some quick calculations and estimate that every week of delay may cost us as much as $1 million in reduced revenue. Activity X is adding three weeks to our critical path and I can reduce that to just one week by adding resources worth $50,000. It seems like this would be a good idea. Do you agree, or are my numbers off somehow?"

But even project managers who have done no upfront optimizing of the schedule and have no data regarding the value/cost of time frequently wind up having to address those things when the project slips and is likely to miss that "deadline" to which everyone in the organization genuflects! In fact, project managers in such organizations are the most likely to have projects that slip past that deadline and therefore are even more likely eventually to need drag and drag cost computation than those who know about them and have used them upfront to optimize the schedule.

Many books and articles have been written about how to recover a slipped schedule; it is a constant challenge for projects and project managers. This recovery is optimized by including drag and drag cost computation. Computing drag and drag costs on a slipped schedule is really no different from doing it upfront during schedule planning. The most important element is simply that on a slipped schedule, we work with each activity's remaining duration rather than original duration.

It is also important that all work is being performed per the CPM logical relationships: out-of-sequence work makes it very difficult to be sure of current remaining durations, current float, and thus current drag.

In addition to any changes in remaining durations, float and drag, the other very important item in analyzing a slipped schedule for recovery is drag cost. Frequently, finishing later than the target date involves a much greater cost than any benefit to be gained by finishing earlier than the target date. For example, getting Softy the Snowboy onto the toy store shelves earlier than Black Friday may give us a slight increase in sales, but every day later during the peak shopping season may result in a 20% reduction in revenues. Or perhaps it would be a 15% reduction for one week late, 20% for the second week, and so on. Suddenly, resourcing decisions that were not justifiable when the drag cost was based on an acceleration premium may now be overwhelmingly the correct decision due to being based on the avoidance of a much greater delay penalty.

This relates to another Daniel Kahneman principle from *Thinking, Fast and Slow*. This one concerns loss aversion: that human beings usually worry more about (and thus tend to work harder to avoid) a loss than they do to realize an equal or even greater amount of gain.[3] In the *Softy the Snowboy* case, the monetized loss from being late really is greater, but even if it were identical with the available gain from being early, accruing delay costs would likely affect the responsible project manager's career significantly more than an identical amount of acceleration premium would benefit: $100 is $100, identical whether viewed as lost or simply not gained. But the human tendency of the project manager to invest less effort in gaining the acceleration premium is yet another reason to avoid the deadline mindset and work instead to expected project profit, the DIPP and the DPI.

Computing critical path drag on a schedule subset

Sometimes, especially in contract work, the customer will suddenly have an urgent need for the component to be delivered earlier than originally planned. For example, this is common in US Department of Defense acquisition programs, where an opportunity may suddenly present itself for earlier integration of several components from different subcontractors within a larger system.

Several years ago, I was teaching a class for such a contractor. When I explained the concept of critical path drag and explored how it could be used, one woman in the class became visibly agitated. I asked her why.

> I was working on a program a couple of years ago where if we had had this concept, it would have saved us huge amounts of time and money. The program had a duration of five years. But within the first year, there

was a component that we were supposed to deliver around the end of October. All of a sudden, sometime in March, we got word from the customer that they needed to accelerate that deliverable by six weeks, to the middle of September. We looked at the schedule, but it didn't help a lot because the deliverable wasn't on the critical path. It had about 400 days of float.

We decided to have a meeting in San Diego to hammer out what we were going to do to pull in the deliverable. We all flew in on a Sunday evening, and most of us had a list of ideas that we felt would work. On Monday morning, we sat in a conference room and discussed our ideas and checked off the ones we thought made sense. About 4:00 in the afternoon, we called in the master scheduler, explained the changes, and had him input them to the schedule. When we ran the schedule, the deliverable date came in: from October 27 to October 26!

We all threw down our pens and went out and had gin and tonics. The next day, we came back and just went through everything we could imagine on the basis of trial and error. Suppose we could get this done in 6 days instead of 10? Darn, we'd only gain 1 day. Okay, I think by doubling the resources, we could get this activity in from 24 days to 12 days. What would that do? Oh, it didn't pull it in at all; it wasn't on the critical path to that deliverable! And basically, that's what we did for the rest of the week until, bit by bit, we did pull the schedule in. But if we had understood drag, we probably wouldn't even have had to go to San Diego. A conference call could have resolved the whole thing and saved us lots of time and money.

She was right, and the problem was also more complicated than she described. The reason is that the target deliverable that needed to be accelerated was not on the program's critical path. And I have since discovered that even experienced schedulers often don't know how to deal with such a situation. Here is how you can address it:

1. Create a subfile and make the delivery of the component the last activity, without any successors.
2. Delete from the subfile any activity that is not an ancestor of that delivery.

3. Identify the critical path among the ancestors and the total float of those that are not critical.
4. Calculate the drag of the critical path activities.
5. Compress the durations and fast-track the critical activities, start-ing with those that have the largest drags, until the critical path goes elsewhere.
6. Then recalculate the drags of the new critical path, rinse and repeat until the delivery's schedule is acceptable.

The overall purpose of this anecdote is to show that, although tradi-tional critical path scheduling and analysis are of great value, the drag adds crucial functionality that aids the project manager with what is frequently one of the hardest parts of the job: schedule recovery. The ability to iso-late the schedule impact of each activity in the remaining schedule greatly facilitates the project manager in performing this key part of the job. By using drag to compress the schedule right up front, during the planning process, the project manager can also create sufficient schedule reserve to make it much less likely that he'll have to perform schedule recovery later.

But when critical path drag is combined with the investment model of the project where the impact of project duration on the expected value is estimated, suddenly the drag cost metric becomes extremely powerful. From this springs the ability to conduct a level of cost–benefit analysis that is almost impossible without it. With drag cost, it becomes crystal clear that the true cost of a critical path activity must be greater than one that is not on the critical path. It also brings into question whether an activity that has a large drag cost is really worth doing. It was easy to decide to add scope, such as an extra bell or a whistle or input field on the product, when the cost of doing so was assumed to be simply the $6,000 overhead-burdened wages of Joe Schmo for two weeks. But suddenly those extra weeks are going to have a drag cost of $100,000 each! Is it still worth including that extra scope? This is where the value breakdown structure, requiring estimates of the value-added of project components, can help us decide. If the value-added of the whistle is only $150,000, then spending $206,000 to include it is clearly a bad investment.

But what if we increased the resources to add the whistle? What if in addition to Joe Schmo, we also added Flo Stowe and Mo Woe, and thus reduced the drag to one week? Now if we assume the cost of the resources has tripled, the true cost is $118,000.

The cost of additional resources to reduce greater amounts of drag cost suddenly can be justified, and the result is greater expected profit (as measured by the DIPP and the DPI) from the project investment. $118,000 to generate $150,000 of value? That might be worth it.

What if the additional resources are not available, either because the functional department's headcount is insufficient or because Flo and

Mo are assigned to another project? Sometimes projects are delayed not just by work, but also by resource constraints. Resource bottlenecks can have drag, too. This is what we discuss in the next chapter, where the investment metrics of projects combined with the new scheduling metrics can give us techniques to determine appropriate staffing levels for each department in the entire project-driven organization.

Summary points

1. Project management is not an easy job, and the analysis techniques described in this book require effort. But the amount of money (and sometimes human lives) riding on a project can be huge. An analytical technique that adds just 1% to the expected monetary value of a medium-sized project with a $10 million budget is worth at least $100,000. Surely it is worthwhile to employ both such a technique and personnel who know how to use it.

2. The value of good management of project investments accrues much more to the organization and to senior management than to the individual project manager or engineer. But the effort of doing good project management rests on the shoulders of the project team. If senior managers do not both mandate and support good project management techniques, they will not be used.

3. There are two different types of work on any project: mandatory and optional. Optional activities often give the project team some wriggle room, but they also can wind up having a true cost that is more than their scope is worth. Estimating the value-added of optional activities, monitoring them to see if they migrate to the critical path and then contrasting the value-added with their true cost can avoid the peril of doing work that costs more than it's worth. This also includes doing a project that delays generation of the program's value. In such a case, the whole project can have drag and drag cost based on the program's schedule and business value.

4. A damage control time chart is a crucial document for analyzing how the expected value of a project may change due to acceleration or delay in computing drag cost at the activity level. On some projects, the damage control time chart measures the cost of time in property damage, injuries, or lives lost. On other projects, it is useful for measuring the potential gain or loss of revenues and other values due to acceleration or delay.

Even if drag and drag cost are not computed during the planning stages of a project, they can be absolutely crucial in helping a project team discover options to recover from schedule slippage.

Endnotes

1. Daniel Kahneman, *Thinking, Fast and Slow*. New York: Farrar, Straus and Giroux, 2011, pp. 44–45.
2. *A Guide to the Project Management Body of Knowledge*, fifth edition. Newtown Square, PA: Project Management Institute, 2013, p. 553.
3. Kahneman, pp. 282–288.

chapter seven

Of resources and rightsizing

"How should the investment model guide an organization's staffing levels?"

How important are projects to any given organization? It varies depending on the type of organization. In many industries (construction, US Department of Defense contractors, product development and marketing including electronics, automotive, medical devices, pharmaceuticals, and many more), projects generate or enable up to 99% of corporate revenues. In other industries (energy generation and distribution, Internet and cable providers, hotel and restaurant chains), day-to-day operations are what is viewed as paying the bills, but these operations too are enabled by critical projects: nuclear plant refueling, oil well and pipeline creation, erection of cellphone towers, residential and commercial construction, and so on. Even industries such as banks, insurance companies, department stores, and hospitals have whole departments like IT that are often more than 90% project-driven and without which the larger corporation would be substantially disabled.

For years I have advised young people looking for a job to work for companies where their particular skill is at the essence of how the company generates revenues. All else being equal, an electrician is likely to feel more appreciation working for an electrical supply company than for an insurance company. This probably shouldn't be the case: the insurance company's electrician undoubtedly does a lot of work over the course of a year that enables the company to generate its revenues. But that enabler factor is usually not fully appreciated. And that is a big part of why project management methods and culture are often so much less appreciated and mature in those giant IT departments of operations-driven (but project-enabled) businesses than in those where projects are recognized as being the life's blood of the corporation.

Whether a project is in a project-driven business or a project-enabled one, planning and managing it in investment terms offers the ability to quantify its value clearly, the value/cost of schedule acceleration or delay, and thus the value of needed project resources and the cost of not getting them. On an IT project designed to provide a mobile app to the sales representatives of an insurance company, if the lack of two additional programmers will delay rollout by two months, that information must be monetized: how much will sales revenues be reduced as a result of the delay? That is the engine that drives the insurance company, and the IT project is an enabler. Without that connection, it's just those software geeks wanting more money to do something that none of the insurance executives really understands anyway.

Getting the needed resources is a battle in every industry and organization, but it is particularly difficult on projects that are not direct drivers of the corporate revenues. Using the techniques in this chapter can completely alter that.

How we got to this point

So let us assume we have performed all the planning processes as recommended in the earlier chapters:

1. The project was initiated from an investment point of view, and its expected monetary value has been estimated as part of the project initiation documentation (e.g., the project charter and business case). This estimate took into account whether the project is an enabler, and if it is, what impact it will have on the expected value of other projects or processes that it is enabling.
2. The business case also provided the project team with an estimate of the impact on the project's EMV (including the impact on the value of any work it's enabling) of finishing earlier or later than a specific target date.
3. A work breakdown structure was then assembled identifying all of the components and the work plan for the project.
4. The WBS was then used as a basis for a value breakdown structure, in which components and work were determined to be either mandatory or optional, and the value added of optional work was estimated.
5. The lowest level of the WBS was then used to generate a detailed list of work activities.
6. Duration estimates and the resources needed to achieve those durations were then plugged into each of the detail activities.
7. A secondary (DRED) duration estimate was obtained from a subject matter expert for each activity, based on how long the activity would take if its level of resources were doubled.
8. The detail activities were then sequenced, and a network logic diagram of the CPM schedule was developed, showing the drag for all critical path activities. The drag amounts for each critical path activity were then combined with the project acceleration value/delay cost estimates to determine the drag cost of each activity.
9. The true cost of each activity was then determined by adding its drag cost to its budget.

10. The expected project profit was then increased through the process of schedule optimization, targeting the activities with the largest drag costs and seeing where the expected project profit could be improved by a combination of:
    a. Fast tracking (i.e., doing more work in parallel)
    b. Crashing the critical path (using the DREDs for guidance and reducing the true cost of appropriate activities by increasing their resource costs while reducing their drag costs by a greater amount)
    c. Eliminating optional activities whose value-added would be less than their true cost.
11. As the critical path changed due to compression and a new path became critical, the same process was performed on that new critical path, and the process iterated until optimization opportunities were exhausted.
12. At this point a "snapshot" of the project's EMV, expected project profit, and starting DIPP should be saved. This represents what we may be able to accomplish if further constraints do not crop up.

Resource availability and project CPM schedule

If we have all the resources we need in sufficient amounts when we need them, then the optimized CPM schedule with an added and considered amount of schedule reserve may represent a reasonable initial target. However, we should be so lucky: almost invariably, activities will have to be delayed because the resources they need are not available on the dates scheduled through CPM.

There are four things that delay activities:

1. *The logic of the work.* These are the logical constraints that are often driven by physical laws: you simply can't debug the software code until after you've written it nor hang the door until you've attached the hinges.
2. *The nature of the work.* Some work is harder, more complex, riskier, or simply takes longer. Additionally, external factors such as climate may combine with the nature of the work to preclude it from occurring at certain times.
3. *Resource availability.* It doesn't matter that the CPM schedule says that the work could occur in August if the resources needed to do it aren't available until September.
4. *Discretionary constraints.* An activity could take place in August, but for reasons of risk or cost reduction, the project manager decides to not do it until September.

There is one cautionary point to make here: it is very important to generate and optimize the CPM schedule before allowing for expected resource insufficiencies and bottlenecks, for two reasons:

1. The CPM scheduling process should be constrained only by factors 1 and 2 above. These are constraining factors that are usually very difficult to change. Yes, with voluminous and ill-defined activities, sometimes decomposition into the separate parts can allow compression through adjusted logical relationships and fast tracking. And sometimes "thinking outside the box" can uncover a faster approach to getting the work of an activity done. But once a schedule has been optimized, further improvement through these methods is both rare and limited. The result is that, in essence, the optimized CPM schedule represents the quickest way to get the work done, extended only by these inflexible constraints. But now this gives us a benchmark for what we can accomplish if there are no other delaying factors and therefore allows us to isolate the amount that factors 3 and 4 are adding to the project duration. If the CPM schedule is, say, eight months, we can now identify the other, less rigid, constraints and see how much longer each threatens to make the project. And because the latter two factors are more flexible, if we know the value/cost of the time that each is adding, it can provide the justification for spending a lesser amount of dollars to resolve the constraint. If a resource bottleneck threatens to add an additional two weeks to the critical path at a cost of $50,000, we can "shop around" for a replacement resource knowing the maximum that it would be worth to our project. And with every such decision, the expected profit of our project investment grows.

2. The second reason for optimizing the CPM schedule before accounting for bottlenecks is that bottlenecks are resolved by delaying work, almost never by accelerating it. Let us imagine that, as shown in Figure 7.1, the resource we need (a hobbit, for instance) is available to our project only during the first week of each month. If we have not optimized the CPM schedule, so that the activity "SOLVE RIDDLE" that requires a hobbit is scheduled for the second week of May, it will be delayed by three weeks, until the first week in June. If this activity is on the critical path, the delay could cause a huge loss in the project's business value. Conversely, Figure 7.2 shows that if we had previously optimized the CPM schedule so that "SOLVE RIDDLE" was pulled in to the last week of April, now it would be scheduled for the May window of hobbit availability, four weeks earlier than with the nonoptimized CPM schedule.

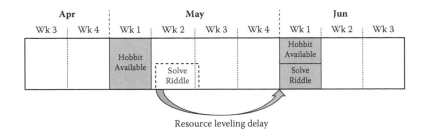

Figure 7.1 Activity resource leveled without prior CPM optimization.

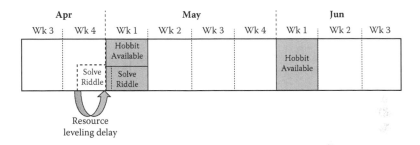

Figure 7.2 Same activity resource leveled after prior CPM optimization.

The additional investment value of the shorter schedule is created almost invisibly, without increased resource costs, all through optimization of the CPM schedule.

Resource unavailability and activity duration estimates

Back in Chapter 4, my recommendation was always to generate duration estimates by assuming a minimum of one dedicated resource of each required type for each activity if assigning less would change its duration estimate. My reasoning was that if we can get 25% of a resource, we can almost always get 100% of that resource: we just need to be able to demonstrate the difference between what we could accomplish (in expected project profit terms) with a dedicated resource versus our limitations due to the restricted resource availability.

Obviously, if we have resource restrictions, we have to take them into account or our working schedule will be hopelessly unrealistic. Now we are in a position to identify the resource restrictions, to measure their effects, and to ameliorate those impacts in the best way possible from the point of view of the project investment. Until we have a CPM schedule, we don't know when the various activities could take place. So even if we know exactly what resources are and are not available in any given time period, we still don't know where unavailability may affect our schedule. This requires the comparison of two different databases:

1. *A database of resource needs for our project, time period by time period.* This comes from having input the resource needs for each activity in the WBS and then scheduling each activity through CPM.
2. *A database of resource availability for our project.* This database is often called a resource library. It contains crucial information for project performance and must be kept up to date as assignments change, staffing levels vary, and employees come and go.

The resource library, and the practice of comparing the project schedules to it in order to identify resource insufficiencies, is the place where many organizations that are otherwise doing a good job of project planning fall flat on their faces. And the destructive results of this failing cause vast inefficiencies throughout the organization.

Role of the functional manager

In most corporations, functional departments are headed by someone who is called a functional manager. The functional manager is typically an experienced subject matter expert in some technical area: programming, documentation writing, mechanical engineering, hobbiting, and so on. He or she needs to have technical expertise for several reasons:

- To staff the department with other technically qualified people
- To make or assist with estimates of the technical resources needed for a given activity
- To assign the people with the right skills to the right projects
- To judge the technical performance of such assigned individuals
- To ensure that a sufficient number of members of the department are trained in those skills that are needed now and those that are likely to be needed in the near future
- To make sure that members of the department have enough work to keep them busy

These are all very important roles whether the organization is 100% project driven or is primarily devoted to operations and only 20% of its working time is spent on projects.

Utilization rate metric

But it is truly terrifying that the item most emphasized in project-driven organizations, especially contractors, is that last one: making sure that the workers in the department are kept busy or, in some industries, billable.

In many places, a functional manager's most important metric, the one by which he is judged for promotions and raises and bonuses, is what is called his department's utilization rate. That is, simply, the percentage of work hours among the staff where they are assigned to projects and thus their time is paid for by the customer.

There is no doubt that, in any project-driven organization, there is value to keeping employees on billable projects or contracts. On cost plus or time and materials contracts, the money that the customer pays per hour for the resource's time is the contractor's revenue. And, as discussed earlier, this can lead to unintended incentives for the contractor to dilate the project so as to keep employees busy and the revenue streaming in. But even on fixed price contracts, the effect of measuring the functional manager on the basis of the department utilization rate is exactly the same principle: he is being judged and rewarded based not on the value added to the project's work but on the number of billable hours his department can charge against the project.

How can a functional manager maximize the number of hours that his department's employees are billable?

- He can deliberately understaff. If he staffs at a level to meet needs when there is great demand for his department's technical skills, then when demand slackens there will be people who don't have a project against which to charge their time. And the department utilization rate will decline. So he will instead be reluctant to hire more people until he is absolutely sure of being able to keep them all very busy. So what if the understaffing results in bottlenecks that delay projects? Well, his job performance isn't being measured on that.

- He can keep his people as multitasked as possible, moving them perhaps several times a week between different activities on different projects. That way there will always be billable work for them to do. So who cares if Henry Hobbit is only assigned as a 25% resource to a critical path activity on a project with potentially huge delay costs, an activity that has 20 days of drag which could be lowered to 5 days of drag just by assigning Henry full time? The functional manager is not being measured on that either.

- When a project manager tells the functional manager that she's going to need Henry full time during the middle two weeks of August, he tells her that if she wants that, then she's going to have to take Henry for the whole month of August. Unless she does, he says, he wouldn't be able to keep Henry billable during the first and last weeks of the month. Or maybe he'll have to assign Henry full time to another project where a different project manager has assured him that he'll keep Henry the whole month. And the second project may go over budget to pay Henry for time when he's not needed.

- As another large project to which Henrietta Hobbit (Henry's sister, of course) has been assigned for the past year starts to wind down, the functional manager announces that Henrietta will have to start dividing her time between the old project and a new one that is just starting up. But couldn't a different resource be assigned to the new project, so that Henrietta's critical path work is neither interrupted nor delayed? Sure, but then what would happen when Henrietta's work actually finishes on the old project? What if there is no new project immediately available for her to which she can be assigned? Then the functional department's utilization rate might go down, and we can't have that happen, can we? "Henrietta, we need you to start spending time on the new project; it doesn't matter that delay costs will be incurred by the old project as a result."

Again, there is that old saying, "Be careful what you measure because you may get it!" As we saw earlier, working to deadlines, when combined with the human behaviors described by Parkinson's law, can result in actions that have negative consequences for the project investment. When we get to discussing earned value and, specifically, the schedule performance index (SPI), we see an important and much used metric in project control that can spur counterproductive actions. The utilization rate metric is another. Fortunately, it is used in only a relatively small subset of organizations, but those that are using it should either abandon it or be very careful how it is applied.

Maintaining the resource library

Among the responsibilities of the functional manager is that of ensuring the resource library database is maintained with the most up-to-date information. Project managers then use this information to "people" the activities in their projects.

- The most basic format in which the resource library resides is in the brain of the functional manager. This format allows the functional manager maximum control over his department's resources by being completely opaque to any and all project managers who are not mind readers. Those who are not are therefore completely at the mercy of the functional manager's willingness to be collaborative and seek ways to help the project. Some functional managers play everything close to the vest and try to make their departments into black holes, from which not even information can escape: "I don't know when I'll be able to give you someone. And I don't know who it will be. Yours isn't the only project I've got to worry about, you know. If someone comes available, I'll let you know. But I can't guarantee anything." And of course, the lack of visibility of resource assignments enables such opacity.

- The second most basic format for resource library information is the matrix drawn on the whiteboard. Across the top of it is a date ribbon displaying the weeks and months out into the future. Listed vertically along the left side are the resources, usually by individual names: Asif, Lindsay, Janet, Paolo, Selena, and Virender. Next to each name is a bar, in Gantt chart form, showing the project each person is assigned to across the dates of the assignment. This format provides more visibility to the project manager who visits the functional manager's office to plead for resources, but control still resides with the functional manager, and she's free to change the rules as she goes along: "Actually, I just found out that Selena's work on Project XYZ might wind up extending for another six weeks. So I really won't have anyone for you until March, when Janet should be available. But that could change too."
- Some functional managers get very high-tech and put the resource library information into an Excel spreadsheet. It's actually more opaque than on the whiteboard, but it seems a lot more businesslike.

The resource library does not belong in any of the above formats: it belongs in the project management software where the project schedules also reside. And failure to do this makes project-driven organizations much less efficient. The resource library needs to be in the same software as the project schedules because there is a two-way street between resource availability and a project schedule, and a multiway street between all of the project schedules and resource availability.

The projects need resources in order to get done, but the timing of that need is uncertain and changeable as the working schedule changes. This uncertainty makes resource assignments more than six months out fruitless, and those in the three-month to six-month range dodgy. But dodgy does not mean worthless: that unpredictability makes it worthwhile to maintain the database and to track at least a general sense of future resource needs.

Resource needs during the next two or three months are also dodgy, but the immediacy of events makes it crucial for projects to be able to track these and adjust their windows of needs if the schedule changes. And although multiple projects sharing the same database of resources makes the task harder, it also makes it even more crucial. An activity that is scheduled for the first two weeks of June may have an ancestor activity scheduled for April that slips by in three weeks. This will change both the availability of the resources on the ancestor activity and the availability of resources on all of its descendants. There may have been nothing that could have been done to stop that slippage in April, but it is crucial that its impact on available resources to the downstream activities be clearly visible as soon as possible.

The more projects that are churning through their schedules and causing changes in resource availability, the more crucial it is to try to see the impact of the churn. And the only way to see that is if the resource library and all the schedules are located in the same project management software on the same server, in other words, a portfolio-wide project management system and resource library. Project-driven organizations that try to save money by sticking to a few dozen desktop software packages, with each project schedule invisible to the others and resource loading (if it is done at all) performed as a result of weekly meetings and manual entry, are engaged in an effort that is penny wise and pound foolish. The whole process could be streamlined and made much less labor-intensive with a server-based system.

- The schedule-based resource needs of each new project, as soon as its initial schedule and resource needs are uploaded, could be projected and matched against the availability within the resource library.
- Bottlenecks (resource over-allocations) can be immediately identified for the next few months, and discussions about how to ameliorate the problems begun.
- And then, as schedules on most of the projects change on a weekly or daily basis, the new bottlenecks caused thereby can be identified, their impacts forecast, some of the new problems addressed easily (such as an activity being forced to slip by two weeks, but it has five weeks of total float) and others needing to be discussed.

Working around lack of a resource library

Project managers know that they will need specific resources at specific times or else activities will be delayed. If any of these activities are on the critical path, the result will probably be later completion, meaning lots of gray hair for the project manager and loss of value for the organization. Each resource bottleneck will hit the project schedule like an iceberg in the dark, often sinking the expected project profit. This happens again and again on project after project in project-driven organizations.

Conscientious project and functional managers, knowing that resource bottlenecks are likely to be both ubiquitous and disastrous, will often try to "work around" them. Unfortunately, this is too frequently a case of the medicine being worse than the cure.

Several years ago, I consulted on a program with a large engineering company where the purpose was to manufacture, test, and ship six almost identical components, in series. Unfortunately, the testing facility was available for only a few days each month and the bottlenecks that this would cause would make each component's delivery later than the contractual date. Addressing the first component, I worked with the project manager using drag, DRED, and true cost metrics, to try to pull in the testing activity to fit into the earliest possible window of availability. Unfortunately, a two-week activity immediately before testing stood in the way: "PREPARE TEST FACILITY."

The program manager and I visited the functional manager responsible for the test preparation, showed him our schedule and how his activity's drag, combined with the small availability windows, was going to make every component late. He shrugged: "I can get preparation done in two days if you give me enough warning."

I stared. "But—but why does it say it's going to take two weeks then?"

"Because I don't want to have my people all sitting around just waiting to work on your program! They're off doing other things for other programs. But if you tell me you need them on this date, and you stick to that, I'll make sure they're available to get it done. But don't let me down!"

I immediately understood three things:

1. In a multiproject environment where every program manager feels empowered to come in two days before he needs a test and demand that the facility be ready, the functional manager takes the only measures available to combat chaos and save his own sanity: he pads the schedule.
2. This functional manager was not the only one in the organization doing exactly this sort of thing. Most functional managers (charged, remember, with maintaining high utilization rates) were protecting themselves by providing estimated durations that allowed them time to gather resources and then do the work, because there was no system that allowed them visibility into the churning program schedules to see how upstream delays would affect the date that they needed to be ready.
3. Even when functional managers might not be causing delays through padding, program managers cannot be sure of that and so might themselves input padding to the CPM schedule, knowing that it might take two weeks for a functional manager to round up the resources needed for a two-day activity. If we're not going to have resource-adjusted schedules, then we have to make sure that our CPM schedules are accounting for the myriad bottlenecks that will surely occur.

The lost contracts and revenues that this behavior undoubtedly causes on scores of proposals and programs are beyond reckoning, and all due to the lack of a reliable multiproject resource library interfacing with the resource needs of all the projects.

Resource leveling

Every project software package that rises above minimal functionality is equipped to perform something called resource leveling. This is actually not one thing but two, based on two different algorithms programmed into the software. The purpose of resource leveling is twofold:

1. To stabilize staffing levels for a project or for a project-driven organization
2. To demonstrate the schedule slippage that can be expected due to bottlenecks

Resource leveling can be performed on a single project within a desktop software package or on a multiproject basis, across all projects and all resources in a portfolio or organization, on a server-based project management system. In a multiproject environment, multiproject leveling offers a great many benefits that single-project leveling does not. But for simplicity's sake, we describe the process as it would occur on a single-project schedule. The implications of the extension of these methods to a multiproject organization have even more impact.

Time-limited resource leveling

All resource leveling is a function of comparing a project's resource needs during different time periods against resource availability across the same time periods, and setting priorities. Some software packages allow the user to define higher priorities for certain activities than for others, meaning that the software algorithm, if faced with a choice of which activity should get the resources first, will assign them to the one with a higher priority. In multiproject leveling, one project can be given a higher priority than another so that resources will only go to the second project if they are not needed at that time on the first project. But these sorts of user-defined priority settings should be a product of the resource leveling function rather than a driver of it. The two fundamental priorities are time and resources:

a. Should the project schedule maintain a fixed completion date even if the resources are not available to achieve it?
b. Should the scheduled completion be delayed in response to identified resource bottlenecks?

The first of these alternatives, the project maintaining a fixed completion date, is a function of what is called time-limited resource leveling. Typically, the software user tells the program that she wants to level the resources but maintain a specific finish date. The software will let the user enter any finish date she wants, but my very strong suggestion is always to start with the CPM finish date. Why? Because that is the finish date that our CPM analysis says we can achieve if there are no resource bottlenecks. (That's why we based our duration estimates on dedicated resources, 100% allocations of each. During this resource leveling process is where we will account for what happens if a given resource is only available to us 25% or 50% of the time.)

In performing time-limited resource leveling, the algorithm will manipulate activities in certain ways in order to get rid of bottlenecks. The user may choose to allow the software to split activities into two or more pieces if doing so will resolve the bottleneck, or not. But the one thing that the software will not do is delay the project beyond the input finish date. And if that date is the finish date of the CPM schedule, then the algorithm will delay no activity beyond its total float. That means that no activity on the critical path can be delayed at all.

After performing time-limited resource leveling, there will still almost always be bottlenecks, resources that the resource library shows are over-allocated. Some software packages are very good at pointing the user to precisely which resources are over-allocated and during what periods; with other packages, the user may have to do a little hunting. Under any circumstances, each type of resource (hobbits, wizards, mechanical engineers, etc.) that is over-allocated and the period of over-allocation should be noted. A histogram displaying the over-allocation of hobbits on a project schedule after the time-limited resource leveling was performed is shown in Figure 7.3.

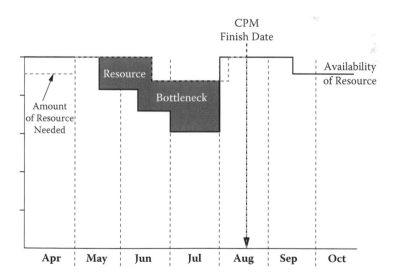

Figure 7.3 Histogram of time-limited schedule of usage to meet CPM finish date.

Resource-limited resource leveling

After we have noted all the bottlenecks that the time-limited resource leveler was unable to resolve without delaying project completion, we should toggle our software to generate a resource-limited resource schedule. In this case we are telling the program to ignore our CPM finish date. We still want to get rid of the bottlenecks, but now the level of availability for each resource is the constraining factor. So the algorithm can delay the end of the project, if necessary, by delaying both critical path activities and noncritical activities beyond their total float.

The software cranks once again and gives us the schedule in which our end date has been delayed, but all of our bottlenecks have been removed by delaying activities to those periods when sufficient resources are available. This is where we see what only having a 50% hobbit and a 25% mechanical engineer will do to our schedule. Once again we should examine, one by one, the resources that were bottlenecked after time-limited resource leveling and see what their impacts are on the schedule. A histogram of the resource usage of hobbits in the resource-limited schedule is shown in Figure 7.4.

One more cautionary word on resource leveling: the robustness of different software algorithms can vary tremendously in their ability to produce a resource-limited schedule that is as short as possible. Some higher-priced packages do an excellent job, cheaper packages, not so much. Under any circumstances, it is a good idea for the user to eyeball the schedule changes that the software algorithm makes and check to see

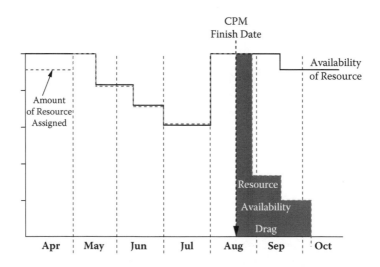

Figure 7.4 Histogram of resource-limited schedule of usage.

if there isn't a better adjustment that could be made that will allow for a shorter schedule within resource availability limits.

But when all adjustments have been made, the project's working schedule should be the resource-limited schedule, the schedule that takes into account the impact of those resource bottlenecks that have not been resolved. It is therefore important to know the impact of such bottlenecks on the project's expected profit, so as perhaps to justify additional resources to make the bottlenecks disappear, on this or future projects.

The cost of leveling with unresolved bottlenecks (the CLUB)

The resource-limited schedule shows that, in order to accommodate the resource insufficiencies, our project completion date has slipped out by a certain amount of time. The majority of that small minority of project managers who even use this functionality will now adjust their schedules to reflect the resource-limited output and start negotiating for more time with the customer/sponsor. The best practice is to use it to negotiate for more of the resources whose bottlenecks are causing the delay(s). Unfortunately, they are often engaging in a losing battle as additional resources always cost money, whereas delays are measured as only costing time. And in the business world, dollars trump days pretty much every time.

The only reliable way to justify the cost of the additional resources is to change days into dollars. And then we're back to where we began this book, with the need to manage projects as investments and to translate time into dollars: what is the cost of the delay, in reduced value-over-cost (or expected project profit), that the lack of 50% more of a hobbit is causing? If the project were a nuclear power plant fueling outage in an energy grid where every day of downtime costs $500,000, and the delay due to a half-time hobbit (although hobbits, admittedly, are often insufficiently utilized on nuclear refueling outages) is two days, we will have justified an expense of anything less than $1 million for getting a full-time hobbit. On this project, the CLUB (i.e., cost of leveling with unresolved bottlenecks) for the hobbit resource is $1 million.

Of course, the resource-limited leveling lumps the bottlenecks of all resources together. To now apply the CLUB metric in order to generate a more profitable schedule, we really need to isolate each bottlenecked resource one by one, and see what impact each is having on our resource-constrained schedule. This process is similar to the optimization of the CPM schedule by using drag and drag cost.

It is likely that, due to one or more bottlenecks, we now have a critical path that is different from our CPM critical path: the bottlenecks have swallowed whatever float there was on this path so that it has now become the longest path and has drag. This drag is different from the CPM drag, which was caused by the logical constraints and the time required to do each work activity. This new drag is resource availability drag (or RAD), caused by specific bottlenecks of specific resources during specific time periods.

Sometimes additional resources are unobtainable. But much of the time they simply require financial justification. If our project has been allocated Henry the Hobbit only half-time, what is Henry doing the rest of the time? If he's assigned to another project, what would the impact be on that other project? It is quite possible that Henry's work on the other project is completely off the critical path with lots of total float. (This, of course, is another way in which multiproject organizations that don't mandate the use of CPM scheduling on all projects are leaving themselves open to gross inefficiencies; it is impossible to prioritize resource assignments to project critical paths if we don't even know where the critical paths are.)

But even if Henry Hobbit is initially assigned to the critical path of another project, we now have a way of judging what the best division of Henry's time would be: not simply which project would be delayed more in days (i.e., the critical path drag) if Henry worked on one rather than the other, but rather what would the cost of that time (i.e., drag cost) be across the portfolio of projects? Most often, this will involve simply comparing drag cost on Project A versus the drag cost on Project B. If Henry's being assigned full-time to Project A results in a resource-availability drag on Project B of 10 days at $20,000 a day, and being assigned full-time to Project B causes a RAD of 5 days at $50,000 a day on Project A, then assigning Henry full-time to Project B is preferable to the tune of $50,000. With these data now clearly visible, other what-ifs can also be performed: what would happen if Henry were assigned to Project B 75% of the time and to Project A 25%? Additionally, sometimes other projects may also be affected by the decision and so those impacts should enter into the analysis.

Ultimately, our resource targeting decisions go from the equivalent of shooting black arrows at an invisible target on a dark night to data-guided missiles, where we not only can see what is likely to be the best decision out of many potential options (recruit more hobbits?), but also what the negative impacts might be of any shortfall. And this comes as a result of combining project management software functionality of CPM and resource leveling (all standard methodologies, even if grossly underutilized) with metrics and techniques that treat projects as the investments they are

- Their expected project profits
- The value/cost of time and schedule changes
- The ability through critical path drag to isolate the impact of specific factors on the expected project profits, both on individual projects and across the program or portfolio

Elusive goal: Stable staffing levels for a project-driven organization

One of the great challenges for project-driven organizations is to maintain staffing levels that are adequate for the needs of the projects, yet to avoid the waste of employees sitting around contributing little or nothing while drawing a paycheck. This is extremely difficult to accomplish because of the volatile nature of each project's resource needs. Unlike operations-type work such as manufacturing or hotel management or telephone support or other enterprises where daily work levels are likely to be relatively stable, project daily work levels are anything but stable. Figure 7.5 models the resource utilization of a typical project, starting with a small team doing planning, rising to a large team working busily every day to perform the project, and then shrinking once again to a small team doing integration testing, final corrections, packaging, and delivery. Obviously, individual projects may vary a bit from this model. Nevertheless, it is representative of the resource usage on many projects. The predecessor–successor logical relationships of critical path scheduling will always cause some work to have to wait until other work is done.

The volatility of work in a project-driven organization is multiplied by the fact that there is not one project with volatile work levels, but many, all with different life cycles, at different stages, and needing different resources. The result is that the resource needs of a project-driven organization are analogous to a mountain range, as shown in Figure 7.6, with summits of resource needs followed by valleys of idle time.

Unfortunately, many of the tactics that are used to try to "calm" the volatility result in damage to the projects and the profit that they generate. These include things we've already discussed: multitasking, high utilization rates, and resource leveling, all of which serve to make projects longer and reduce their value. And of course, because the cost of

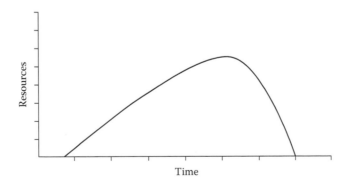

Figure 7.5 Resource usage across the life cycle of a project.

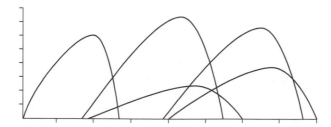

Figure 7.6 "Mountain range" of projects needing resources in a project-driven organization.

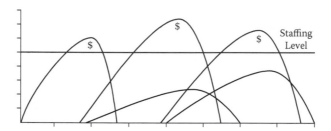

Figure 7.7 "Mountain top" removal through setting of arbitrary staffing levels.

time in almost every organization remains an unquantified externality but the cost of extra resources is clear, we are unable even to see, far less monetize, the destructive nature of these practices as they attempt to tie projects down on the Procrustean bed of stable staffing levels. Resource usage levels are drawn arbitrarily at a certain height up each "mountain," thus "economizing" by reducing the cost of additional resources, as shown in Figure 7.7.

But as the summit of each project is bulldozed off into the next valley as shown in Figure 7.8, the work is always shifted out, to the right-hand side of the schedule, potentially making every project later. And that can result in huge delay costs, on project after project after project. And, of course, delaying work on one project shifts the resource usage to the right, which reduces the resources available to other projects and can delay them as well. The entire portfolio schedule and resource library churns out the delays.

Volatility in resource usage rates is the nature of project work, and nothing is going to change that. If we don't like working in an environment with volatile resource needs, we should not work in project-driven companies. Unfortunately, many organizations, and the human resources departments within them, seek to have a stable staffing level as though it were some sort of Holy Grail. Due to the very essence of a project, this is never going to happen in a project-driven environment.

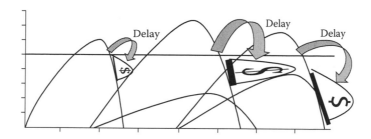

Figure 7.8 Delay costs caused by arbitrary staffing levels.

The sooner we recognize the volatility as an inescapable part of executing projects efficiently and seek to manage it in the optimum way, the better off we will be. The only way of reducing the volatility is to delay work, and if that work is on the critical path, it will reduce project profit. This will become clear once we start quantifying the value/cost of time that has to be traded off to get more stable staffing levels. It doesn't mean that maintaining stable staffing levels has zero value; it's just that we need to know the cost, and that cost comes in the form of longer schedules due to insufficiencies of specific resources on specific activities at specific times, and their cumulative drag costs.

Rightsizing staffing levels for a project-driven organization

Every project-driven organization with which I have ever consulted has had gross resource shortages in specific skill areas that wind up on the critical path of project after project, year in and year out, delaying project completion and reducing project value. One would think that after a while these recurring bottlenecks would get addressed. But they don't. Why not?

1. The organization may have tried, but found it impossible to acquire more of specific resources.
2. The organization may not be using critical path scheduling.
3. The organization may not be using an enterprise project management software system on the server with a multiproject resource library that provides visibility of bottlenecks affecting critical paths.
4. The project managers do their estimating around the bottlenecks on key resources that they know will be there, and so those bottlenecks and their impacts are never identified and never clearly measured.
5. The functional managers are being measured on their departments' utilization rates and easing the critical bottleneck situation would be to their detriment.

6. The functional managers would be happy to hire more critical resources, but the data they need to justify the new hires reside with the project managers and are not being communicated in monetized terms to the functional managers. ("Monetized terms" translates to RADs and the CLUBs for specific resources.)

The first explanation, inability to obtain people with the requisite skills, does happen, but is quite rare. Notice that it is somewhat different from "inability to obtain people with the requisite skills for what we want to pay them." If the resource doesn't exist, there is nothing you can do about it. But getting the additional resources is frequently obstructed by such misguided intentions as, "We don't want to throw our salary structure out of whack." If that's the reason, then consider giving the rest of the department raises. But when dozens of weeks and millions of dollars are being lost every year (and sometimes every quarter) by not having one or two more skilled resources, surely this is nonsensical.

The other five explanations result from failure, at one level or another, to utilize the methods described in this book. Entering CPM schedules (where activity durations have been estimated on the basis of full-time project assignments) into a multiproject resource library will enable bottleneck identification and the measurement in time units of the delays caused by the specific bottlenecks of resources. If the value/cost of time on each project has been identified, not only will it be possible to target the resources in the best way, but the cost of those delays on all the projects can be summed to specific skills and specific departments. Now the only remaining step to justify new hires is communicating the cost of each CLUB to the functional managers, to the recruiting section of the human resources department, and to senior management.

Of course, it may be financially impossible to go out and hire all the resources that are delaying the critical paths of projects. There may have to be some prioritizing of needs. This should be managed by the human resources department (if it's human resources we are talking about, otherwise, departments in charge of equipment acquisition or inventory should perform a similar function). At regular intervals, every three months or so, the department should work with the functional managers to assemble a Pareto chart as shown in Figure 7.9, displaying each resource, from hobbits to mechanical engineers, in descending order according to their CLUBs: how much the lack of availability of the resource has cost the projects in terms of resource availability drag cost. Again, if all the steps we have discussed up to this point are being done throughout the organization, these data should be relatively easy to assemble. The organization can then address the 20% of resources

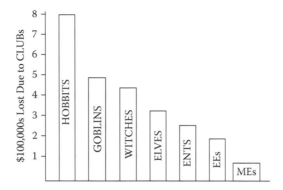

Figure 7.9 Pareto chart of CLUBs of different resources over a previous time period.

that had the highest CLUBs during the previous quarter and begin a recruitment process to lower that cost for the next quarter.

The criticism can be made that looking to improve the future by analyzing data from the past is like driving a car while looking in the rearview mirror. But at least the rearview mirror gives you some information; it's better than driving blindfolded, which is what the vast majority of organizations are currently doing. And just because the data say that the lack of three hobbits, two goblins, and a warlock during the past quarter cost the organization $10 million in project delays does not mean we necessarily have to hire precisely those resources in precisely those quantities. We should, of course, examine the projects that are in the pipeline; perhaps they will need no goblins at all and we should completely switch metaphors and instead hire a genie in a lamp. But at least now we have data to drive the analysis of our functional department staffing levels.

Summary points

1. Obtaining needed resources is always a battle for a project manager, particularly on projects that are not direct drivers of the corporate revenues. This process can be made easier by demonstrating the value of the project (especially if it is an enabler project for all the work that does drive corporate revenues). The key is to focus on and quantify the difference that the added resources will make in the work they enable in terms of expected project profit.

2. The earlier chapters showed how to determine the project's need for resources by creating a work breakdown structure to define scope, and then used critical path analysis to estimate and schedule each work activity and its need for resources. Because resource bottlenecks

are almost always resolved by delaying activities, optimizing the CPM schedule before accounting for periods of resource shortage can result in shorter schedules by planning to do the work before the bottleneck ever occurs.

3. It is important to distinguish between delays that are caused by the nature of work and delays that are caused by resource insufficiencies. It is usually much easier to rectify the latter than the former.

4. Emphasis on maintaining high utilization rates can result in headcounts that are insufficient for optimum project performance. By tracking the CLUBs for specific resources, the impact of such insufficiencies on business value can be measured and the need to increase headcount justified.

5. In an organization with multiple projects, churning schedules often compete for the same resources. It is vital that resource libraries be maintained that will not only reflect current and future resource allocation, but that will change as schedules change, slipping to cause bottlenecks where none existed last week. Such a resource library should be maintained in the project management software on the organization's server where there is multiway communication between the individual project schedules and the resource availability as reflected in the library.

6. Resource leveling to produce a resource-limited schedule can show the impact of a resource bottleneck on expected project profit if the value/cost of time is measured. This information can be used to justify added resources, to increase headcount for critical resources, and to increase the expected value of all the projects in the portfolio.

7. The volatile and calendar-driven nature of a project's resource needs is inimical to the desire for stable staffing levels. Organizations that use resource leveling simply to maintain stable staffing levels are institutionalizing delays and reducing the value of their projects, whether they recognize it or not. A compromise must be reached between the undoubted value of stability and the reduced value of projects due to longer durations. An important element in the resolution of this compromise is an organization's willingness to allow both downtime and comp time.

chapter eight

Fundamentals of earned value

"What is earned value's value if it's cost, not value?"

A quarter century ago when I began teaching and consulting in project management, earned value tracking was a technique used by the US Department of Defense and its contractors and by almost no one else. That situation has changed considerably as earned value management is now being used in lots of organizations, and calculations that use earned value formulas are a big part of PMI's PMP certification exam. (Note that earned value management is EVM and not EMV, which is expected monetary value!)

Unfortunately, whereas DoD contractors understood earned value well (some would say too well as they often showed that they knew how to "game" earned value metrics), the same cannot be said for many of the newcomers to the methodology. There are many misconceptions and many overly ambitious claims that try to make earned value management into some kind of panacea for organizations with troubled projects. And it's not.

But it can be a valuable tool for what it was designed to do, namely, cost trend analysis. It's when efforts are made to force earned value analysis to provide insight into things that it's not designed to do that one runs into trouble.

In this chapter we look at what earned value is and isn't, point out its shortcomings, and suggest improvements that can make it into a somewhat better tool than the way it's currently being used in many places. Earned value is a quantitative technique and so, as in the chapter on critical path scheduling, we have no choice but to resort to formulas and numbers. I attempt wherever possible to keep the numbers simple.

What earned value is and isn't

The simplest and most important concept to recognize about earned value is that, despite its name, it is not about value: it's about cost. Anyone who does not understand the difference between value and cost is hereby invited to a poker game. Value is what you get out of an investment, whereas cost is what you have to invest to get that value.

Both value and cost can be measured in monetary units such as dollars. But there is an important difference in the way that value and cost add up.

Every dollar of cost is like every other dollar of cost: spend $100 on dinner and another $100 on tickets to a show and you have spent $200. The value you get out of the money invested on both dinner and the tickets may far exceed what you would have received had you bought only dinner or only the tickets.

This concept is the kindled value we talked about earlier, where the value impact of the total investment may be greater than the sum of each. This is frequently the case in a program consisting of several projects. You may recall from the earlier chapter that our project to sell Softy the Snowboy toys was timed to coincide with the opening of the *Softy the Snowboy* movie because the Softy toys would kindle the movie ticket sales and the movie would kindle the sale of the Softy toys. Each project by itself would have resulted in less revenue without the other project. This is of course precisely the reason that separate projects should be managed together, as a program. (And it's also the reason that a value breakdown structure works differently from a work breakdown structure, where the costs of each activity can simply be summed to get the total project budget.)

Despite this very important distinction between value and cost, it is surprising how many professionals in the project management field do not understand it. And I believe that this failure to see the distinction led a few years ago to an important earned value metric being given the unfortunate name of *planned value* (PV). Just as earned value is about cost and not value, planned value is really planned cost, based on the budgets for the activities.

Another useful clarification: earned value management is not a project management methodology at all, but a *project control* methodology. What I mean is that earned value data invariably lag behind the work that is being performed, sometimes by as much as one month or even two. If a project manager has to rely on earned value metrics to know that the project is in trouble—that it is over budget or behind schedule—then that project manager has a much bigger problem than the earned value reports would suggest! She is obviously out of touch with the day-to-day work of the project. She needs to know the precise condition of every part of her project long before the earned value reports come out of the finance system.

The earned value methodology was developed in the 1960s by the US Department of Defense when it got tired of giving contractors tens and hundreds of millions of dollars to develop a new weapons system and then would return several months later to discover that all the money was spent but the project was only half completed. There must be some way, thought the DoD, to better control contractor spending and to ensure that the work that was being paid for was actually being done. In other words, before they kept giving the contractor more money, they wanted to make sure that it has "earned the value" of the money they have already provided, and thus the term. Some point out that earned value doesn't have to be based on cost; it could be based on labor hours, or labor dollars or even on expected value. Labor hours, yes, labor dollars, yes, and we discuss the potential value of doing these. But it can't be based on expected value because expected value is not simply summable and earned value is.

Just what is earned value?

Many organizations and project managers shy away from earned value techniques because, simply, they are intimidated by them; they don't fully understand all those acronyms and haven't internalized the formulas. There may be good reason for not using earned value on specific projects, but certainly not because it's too intimidating. In fact, earned value is to project management almost exactly the same as par is to golf. Now, I know many par five holes, and even more par fours, that are extremely intimidating. But I don't know of anyone who doesn't play golf because they don't understand how the par system works.

Just as each hole on a golf course is "weighted" by an estimate of the number of strokes it should require to complete it (if you are a really good golfer), earned value weights each activity by the amount of resource usage or cost it should take to complete it. Add up the par for each hole on the course and you get the total number of strokes it should take to play the whole course. Add up the earned value for each activity and you get the budget (in cost or in use of any specified resource) for the project.

Imagine that I am headed out to play the first three holes of a new 18-hole par 72 golf course. Someone asks me what score I expect to shoot. I of course reply that I expect to play the whole round in 18 strokes. Now you may think that this is highly optimistic, to say the least. But I actually believe that I am being pessimistic; after all, if I can put enough topspin on one of my drives, maybe I can get it to go into the hole, bounce out, roll down the next fairway onto the green and into the next hole. You may be skeptical of my ability to do this, but, believe me, there are project managers who overestimate what they can accomplish by just about as much!

After I play the first three holes, my scores on each hole are 5, 6, and 4. My total for the first three holes, or 1/6th of the course, is 15 strokes. Now if you ask me what I expect to shoot for the entire round, I will of course reply: "Thirty. I had some bad luck on the first three holes but I should be able to get holes-in-one the rest of the way." And there are project managers who are like that, too.

But now we have actuals, data that can be used to update our estimates. If I have taken 15 strokes to play three holes and I have 15 more holes to play, the simplest arithmetic would indicate that my "strokes estimate-to-complete" is another 75 for a total score for the round of 90.

However, if we know what par was for each of those first three holes we should be able to make an even more accurate prediction. If par was 4, 3, and 3, I am playing at 50% over par. For the par 72 round, that suggests a score of 150% of 72, or 108.

If par for the first three holes was 4, 5, and 4, we can estimate that my final score will be 72 * (15/13) or 83 strokes. Simplest trend analysis would have led us to an estimate of 90. But knowing precise actuals-versus-estimates for 16.7% of the effort, and knowing what the estimates are for the remaining effort, allows us to make predictions that are likely to be much more accurate. And this is based on data that are independent of the golfer's own (now clearly optimistic) estimates.

Earned value analysis works in exactly the same way: if the project was budgeted for $7.2 million of which the first three phases were budgeted for $400,000, $300,000, and $300,000, respectively, and their actual costs were $500,000, $600,000, and $400,000, we have reason to estimate not only that we will finish over budget but that our cost estimate-at-completion will be 50% over budget or $10.8 million. We explore the fundamentals of this analysis in greater detail momentarily and perhaps determine ways to make it even more accurate. But just as we have described, it's a useful tool for estimating cost trends.

However, before we leave the golf metaphor, let us ask a different question: if I shot five-over-par for the first three holes whose par added to 10, how long will it take me to play all 18 holes? Now, there is a correlation between the number of strokes that a player requires for all 18 holes and the amount of time it takes to complete the round, but it is at best a weak correlation. Other factors such as how crowded the course is, am I playing alone or in a foursome, am I walking or using a golf cart, and how slowly is the group immediately in front of me playing, will combine to have a much greater impact on the time it takes me than the number of strokes I personally have to play.

If I want to be able to estimate how long it will take me to play the 18 holes, then:

- I need to know what a "par time" is for each hole.
- I need to know how long it took me to play each hole (including the time spent moving from one hole to the next tee).

In golf, time and strokes used are two very different measurements and are weakly correlated. Time and cost on a project are also only weakly correlated. As we have discussed, time on a project is driven by the critical path. Yet traditional earned value management attempts to evaluate and track time on the basis of cost metrics. And, not surprisingly, this does not work any better than tracking time in the golf metaphor. We explore this further.

Fundamental basis of earned value tracking

Par for each hole on the golf course provides a detail-level approach to judging the golfer's progress as he plays each hole. It also allows trend analysis so that we can make an accurate prediction of his final score by assessing his score versus par after three or four holes. We then incorporate the score on each additional hole and, using the fundamentals of Bayesian probability theory, our prediction should become more and more accurate.

This is fundamentally the same as what is done in earned value analysis. Instead of simply saying that a project has a $1 million budget and then getting to the end and realizing that we have actually spent $1.5 million, we break the budget down to what we expect to spend activity by activity and work package by work package.

If our project with the $1 million budget consists of 100 work packages, that would mean an average budget per work package of $10,000. If we have completed 10 of these packages and have spent $90,000, we might be tempted to assume that we will finish under budget. We have completed 10% of the work packages and so a simple cost estimate-at-completion would project a $900,000 cost. However, it may be that the first 10% of work packages were the equivalent of par-three holes, work that was expected to cost less. Let us assume that the budgets and actual costs for those 10 packages is as displayed in Figure 8.1.

Work Pkg.	Budget	Cum. Budget Total (PV)	Actual Cost	Cum. Actual Total (AC)	Cum. Over/Under Budget
001	$7,500	$7,500	$10,000	$10,000	$2,500 Over
002	$5,000	$12,500	$5,500	$15,500	$3,000 Over
003	$10,000	$22,500	$8,000	$23,500	$1,000 Over
004	$12,000	$34,500	$12,000	$35,500	$1,000 Over
005	$6,000	$40,500	$6,500	$42,000	$1,500 Over
006	$12,000	$52,500	$13,000	$55,000	$2,500 Over
007	$3,000	$55,500	$6,000	$61,000	$5,500 Over
008	$6,000	$61,500	$7,000	$68,000	$6,500 Over
009	$10,000	$71,500	$16,000	$84,000	$12,500 Over
010	$3,500	$75,000	$6,000	$90,000	$15,000 Over
Total	$75,000	$75,000	$90,000	$90,000	$15,000 Over

Figure 8.1 Table showing budgeted and actual costs for first 10 work packages.

Immediately we can see that we are not under budget at all. Based on what was budgeted for these 10 work packages and what we've actually spent to accomplish them, we are $15,000, or 20%, over budget. If this is the way we perform on the remaining 90 work packages that are budgeted for the remaining $925,000, how much would we expect them to actually cost? On the basis of these data, we can project that if current trends continue, our cost overrun on the remaining budget will be 20% of $925,000, or $185,000. This means that our cost estimate-to-complete (cost ETC) is $925,000 + $185,000, or $1,110,000. Our cost estimate-at-completion (cost EAC) is $90,000 that we have already spent plus the cost ETC of $1,110,000, or $1,200,000 for a total cost overrun of $200,000.

If you understand what we have just done, then you understand the basics of earned value analysis. We can (and will) slice and dice these data in different ways:

- We can use the numbers in different formulas to gauge and predict other items.
- We can see if the work packages that have been overspending the most have common traits (e.g., the same contractor or the same functional department) that might suggest we should only expect those future work packages that have the same trait to overspend by as much.
- We can see when the work packages were scheduled to be accomplished and when they were actually accomplished and try to extrapolate (with questionable success) how we are doing in terms of schedule.

But the simple calculations that we have done thus far are the essence of earned value analysis, and cost trend predictions are what it is designed to do best.

Basic earned value cost tracking formulas

Notice that none of what we have done thus far has any connection whatever to the schedule. When dealing with cost, we don't care when the work package was scheduled, when the cost was accrued, or whether the work was done early or late. All that matters is (1) what it was budgeted for and (2) what it cost. These two data items, as shown in Figure 8.1, are called *earned value* (EV) and *actual cost* (AC). And as we said earlier and now can clearly be seen, planned value is really planned cost.

These terms are actually new names for the original earned value terms that date back to the 1960s and are still standard in US Defense Department contracting. The original names, although longer and harder to remember, are better descriptors of what they really represent, but the new and shorter terms have spread to common use outside the Defense Department:

1. Earned value (EV), which the US DoD still calls the budgeted cost for work performed (BCWP), is what was budgeted for the work packages that we have completed.
2. Actual cost (AC), which the US DoD still calls actual cost for work performed (ACWP), is what it actually cost to complete the work that we have performed.

These terms are showing us that although $90,000 has actually been spent, we have only "earned the value" that we thought we would get from spending $75,000. The two standard earned value calculations are *cost variance* (CV) and the *cost performance index* (CPI).

Cost variance is simply the difference between what was budgeted for the work performed and what it actually cost. The formula is:

$$CV = EV - AC \quad (CV = BCWP - ACWP)$$

In our project above with the $1 million budget, our EV is $75,000 and our AC is $90,000, which gives us a CV of –$15,000. A negative CV means that, thus far, our project is running over budget. And notice that the second word in each case is "cost."

The cost performance index is what is used to perform trend analysis, to compute, if current trends continue, what we will have spent at the end of the project. As we saw above, it is what will happen if we continue to get the same level of productivity for every dollar spent for the rest of the project. The formula is:

$$CPI = EV \div AC \quad (CPI = BCWP \div ACWP)$$

In our project above, with our EV at $75,000 and our AC at $90,000, we have a CPI of $75,000 ÷ $90,000, or 0.83. This shows that for every one dollar we have spent thus far, we have gotten 83 cents worth of work performed. If this trend continues, how much will it cost to complete the project? The formula is:

Cost estimate-at-completion (cost EAC) = budget at completion (BAC) ÷ CPI

In our project, cost EAC equals $1,000,000 ÷ 0.83 = $1,204,819. (The CPI is usually rounded to two decimal places. In this case, it's actually 0.833333. If we divided by that number we would see that the cost EAC is actually $1,200,000, which of course is exactly what we got when we did the calculation just using common sense and without the "formal" earned value formulas.)

From earlier chapters, it will be remembered that the formula for calculating the DIPP is:

DIPP = (EMV ± acceleration premium or delay cost) ÷ cost ETC

The CPI is one of the best ways to estimate the cost EAC, and thus the cost ETC:

Cost estimate-to-complete (cost ETC) = cost EAC − actual cost (AC)

In the project above, with a projected cost EAC of $1,204,819 and an AC of $90,000, the cost ETC is $1,114,819 (or $1,110,000 using the unrounded CPI), which again is the same number we got without the earned value formulas.

Thus far, we have explored basic earned value cost tracking methods. In a few pages, we expand these further, making them even better predictors. But first let's explore how earned value is also used in an attempt to track schedule.

Basic earned value schedule tracking

Now we get to where earned value tracking becomes a little funky, attempting to do something that it is really not suited to do: track schedules. Each of the activities and work packages, of course, was scheduled to occur at certain dates during our project. We would hope that that scheduling process was conducted as we have suggested in this book, using CPM and optimization, and then resource loading and leveling. Under any circumstances, we wind up with calendar dates when each item of work is scheduled to occur. The assumption is also made that resource usage occurs when the work occurs and that the cost of that resource usage accrues at the same time. Of course, this is just an assumption; we may not actually cut the check to pay for those resources until weeks or even months later. This can lead to a trap called the *disbursement delay distortion*, where we plan the cost to occur when the activity occurs but do not charge the actual cost until the check gets cut. It is very important for project control that the cost for work is charged as actual cost as soon as the work occurs.

Figure 8.2 shows the project at the end of Week 20, and shows us what work packages were scheduled to be finished at that point, which have been completed and which haven't. It also shows us what has been spent on the work packages that have been started but not yet completed. Even though work packages 011 through 013 were scheduled to have been completed by now, they haven't been. As a result, they add nothing to our earned value. Yet they have accumulated cost, a total of $20,000, making our AC $110,000. Now our cost variance and cost performance index are even worse:

CV = EV − AC = $75,000 − $110,000 = − $35,000

CPI = EV ÷ AC = $75,000 ÷ $110,000 = 0.68

Work Pkg.	Budget	Cum. Budget Total (PV)	Actual Cost	Cum. Actual Total (AC)	Week Completion Scheduled	Week Completed	Cum. Earned Value (EV)
001	$7,500	$7,500	$10,000	$10,000	3	3	S7,500
002	$5,000	$12,500	$5,500	$15,500	5	5	$12,500
003	$10,000	$22,500	$8,000	$23,500	6	6	$22,500
004	$12,000	$34,500	$12,000	$35,500	6	7	$34,500
005	$6,000	$40,500	$6,500	$42,000	8	9	$40,500
006	$12,000	$52,500	$13,000	$55,000	11	11	$52,500
007	$3,000	$55,500	$6,000	$61,000	12	14	$55,500
008	$6,000	$61,500	$7,000	$68,000	13	15	$61,500
009	$10,000	$71,500	$16,000	$84,000	15	18	$71,500
010	$3,500	$75,000	$6,000	$90,000	16	20	$75,000
011	$6,000	$81,000	$5,000	$95,000	19	Not Complete	$0
012	$10,000	$91,000	$8,000	$103,000	19	Not Complete	$0
013	$9,000	$100,000	$7,000	$110,000	20	Not Complete	$0
Total (Wk 20)	$100,000	$100,000	$110,000	$110,000			$75,000

Figure 8.2 Progress report at end of Week 20 showing actuals.

Now, based on our CPI and our budget-at-completion, our cost estimate-at-completion can be computed:

$$\text{Cost EAC} = \text{BAC} \div \text{CPI} = \$1{,}000{,}000 \div 0.68 = \$1{,}470{,}588$$

This projection, when combined with what we've already spent (i.e., our actual cost), gives us our cost estimate-to-complete:

$$\text{Cost ETC} = \text{cost EAC} - \text{AC} = \$1{,}470{,}588 - \$110{,}000 = \$1{,}360{,}588$$

Our project is trending to be way over budget. But now we attempt to extend these numbers in order to reach conclusions about schedule. Through the end of Week 20, we expected to have spent $100,000. This is our *planned cost accrual schedule*, summing the amount of money we budgeted to accomplish all the work scheduled to be accomplished through the end of Week 20. So at the end of Week 20, we had also expected to have completed $100,000 worth of work. In addition to being our planned cost accrual schedule, this is also what is called our planned value or, in the original US Department of Defense terminology, budgeted cost for work scheduled. This is our third and final earned value function. The three, in both their US DoD and new terminologies are

1. Earned value (EV) which the US DoD still calls the budgeted cost for work performed (BCWP), is what was budgeted for the work packages that we have completed.

2. Actual cost (AC), which the US DoD still calls actual cost for work performed (ACWP), is what it actually cost to complete the work that we have performed.
3. Planned value (PV), which the US DoD still calls budgeted cost for work scheduled (BCWS), is what was budgeted for the work we have completed thus far.

At the end of Week 20, we were scheduled to have completed activities 001 through 013, budgeted for (or weighted at) $100,000. This is our PV at the end of Week 20.

We have actually completed only activities 001 through 010, budgeted for $75,000. This is our EV.

The two earned value metrics that are commonly used for schedule analysis are schedule variance (SV) and the schedule performance index (SPI).

Schedule variance is simply the difference between what was budgeted for the work that has actually been completed (EV) and what was budgeted for the work that was scheduled to have been completed thus far (PV). The formula is:

$$SV = EV - PV \quad (SV = BCWP - BCWS)$$

In our project above with the $1 million budget, our EV is $75,000 and our PV is $100,000, which gives us a CV of –$25,000. A negative SV means that, thus far, our project is running behind schedule.

The schedule performance index is what is used to perform schedule trend analysis, to compute, if current trends continue, how long it will take to reach the end of the project. As we saw above, it is what will happen if we continue to work at the same speed for the rest of the project. The formula is:

$$SPI = EV \div PV \quad (SPI = BCWP \div BCWS)$$

In our project above with our EV at $75,000 and our AC at $100,000, these calculations show:

$$SPI = \$75,000 \div \$100,000 = 0.75$$

This says that for every one dollar we have spent thus far, we have gotten 75 cents of the scheduled work completed.

Based on this and our original planned duration of 100 weeks, we can now calculate our duration at completion:

$$\text{Duration at completion} = \text{planned duration} \div SPI = 100 \text{ weeks} \div 0.75$$
$$= 133.3 \text{ weeks}$$

Flaws in earned value schedule tracking

There are many problems with earned value schedule tracking. The most obvious of these is similar to the problem in our golf metaphor when we tried to make projections about the amount of time it would take to play the round of golf based on a par system that measured number of strokes. Attempting to make project duration estimates based on an earned value system that measures cost (or other resource usage factors) just doesn't work very well.

If, as we have said several times in this book, cost and value are two very different things whose quantification obeys different rules, then boy, are cost and schedule ever two different things! Let's take the example shown in Figure 8.3. Here we have added two more activities: 014 and 015. They were not due to be completed until Weeks 22 and 23, respectively. But even though we are only at the end of Week 20, both have been completed: $2,000 under budget for Activity 014 and $2,000 over budget for activity 015.

The work in these two activities is finished; we don't have to pay any more money for it. It therefore makes perfect sense from a cost perspective that the project should receive credit for finishing these two activities. Indeed, both AC and EV columns have been updated by a total of $25,000 to

Work Pkg.	Budget	Cum. Budget Total (PV)	Actual Cost	Cum. Actual Total (AC)	Week Completion Scheduled	Week Completed	Cum. Earned Value (EV)
001	$7,500	$7,500	$10,000	$10,000	3	3	$7,500
002	$5,000	$12,500	$5,500	$15,500	5	5	$12,500
003	$10,000	$22,500	$8,000	$23,500	6	6	$22,500
004	$12,000	$34,500	$12,000	$35,500	6	7	$34,500
005	$6,000	$40,500	$6,500	$42,000	8	9	$40,500
006	$12,000	$52,500	$13,000	$55,000	11	11	$52,500
007	$3,000	$55,500	$6,000	$61,000	12	14	$55,500
008	$6,000	$61,500	$7,000	$68,000	13	15	$61,500
009	$10,000	$71,500	$16,000	$84,000	15	18	$71,500
010	$3,500	$75,000	$6,000	$90,000	16	20	$75,000
011	$6,000	$81,000	$5,000	$95,000	19	Not Complete	$0
012	$10,000	$91,000	$8,000	$103,000	19	Not Complete	$0
013	$9,000	$100,000	$7,000	$110,000	20	Not Complete	$0
014	$10,000	$100,000	$8,000	$118,000	22	20	$10,000
015	$15,000	$100,000	$17,000	$135,000	23	20	$15,000
Total (Wk 20)	$100,000	$100,000	$135,000	$135,000			$100,000

Figure 8.3 Progress report at end of Week 20 showing out-of-sequence work.

include the performance of this work, to $135,000 and $100,000, respectively. If we now compute our CPI and projections, they will have improved:

$$CPI = EV \div AC = \$100,000 \div \$135,000 = 0.74$$

Now, based on our new CPI, a new cost estimate-at-completion and estimate-to-complete can be computed:

$$Cost\ EAC = BAC \div CPI = \$1,000,000 \div 0.74 = \$1,351,351$$

$$Cost\ ETC = cost\ EAC - AC = \$1,351,351 - \$135,000 = \$1,216,351$$

All of this thus far is fine. But now look at the PV column. Because it is only the end of Week 20 and Activities 014 and 015 are not due to be completed until Weeks 22 and 23, respectively, they are not included in planned value (i.e., the earned value baseline) for another two or three weeks. Yet, although those activities' budgets are not added to the PV, they are added to the EV column. And the result is that our SPI and duration at completion are markedly improved:

$$SPI = EV \div PV = \$100,000 \div \$100,000 = 1.00$$

$$Duration\ at\ completion = planned\ duration \div SPI = 100\ weeks \div 1.00$$
$$= 100\ weeks$$

Wow! We're back on schedule! But are we really? Let us think back to what determines the duration of a project: the length of the critical path. Yet earned value metrics pay absolutely no attention to critical path; an activity that's off the critical path and has 100 days of total float and a budget of $40,000 has exactly four times the earned value "weight " of an activity on the critical path with 20 days of drag but a budget of only $10,000. What if a project's SPI is below 1.00, but the amount by which it's below is all due to activities that are (1) off the critical path, (2) have slipped a bit, but (3) still have float? In that case, our project's critical path activities all could be on schedule or even early, and yet, as measured by SPI, our project could seem to be running late. In other words, the SPI metric can be distorted by total float.

Let us take as examples the two projects shown in Figure 8.4. Our data date in each case is the end of July, at which time our earned value baseline or PV indicated we should have accrued $5,500 worth of earned value. The checkmarks show us the activities that have been completed. In Project A, the earned value is $4,100 for an SPI of 0.75. In Project B, the earned value is $5,200 for a much higher SPI of 0.95. A quick comparison would tell us that Project A is seriously behind schedule but Project B only slightly so.

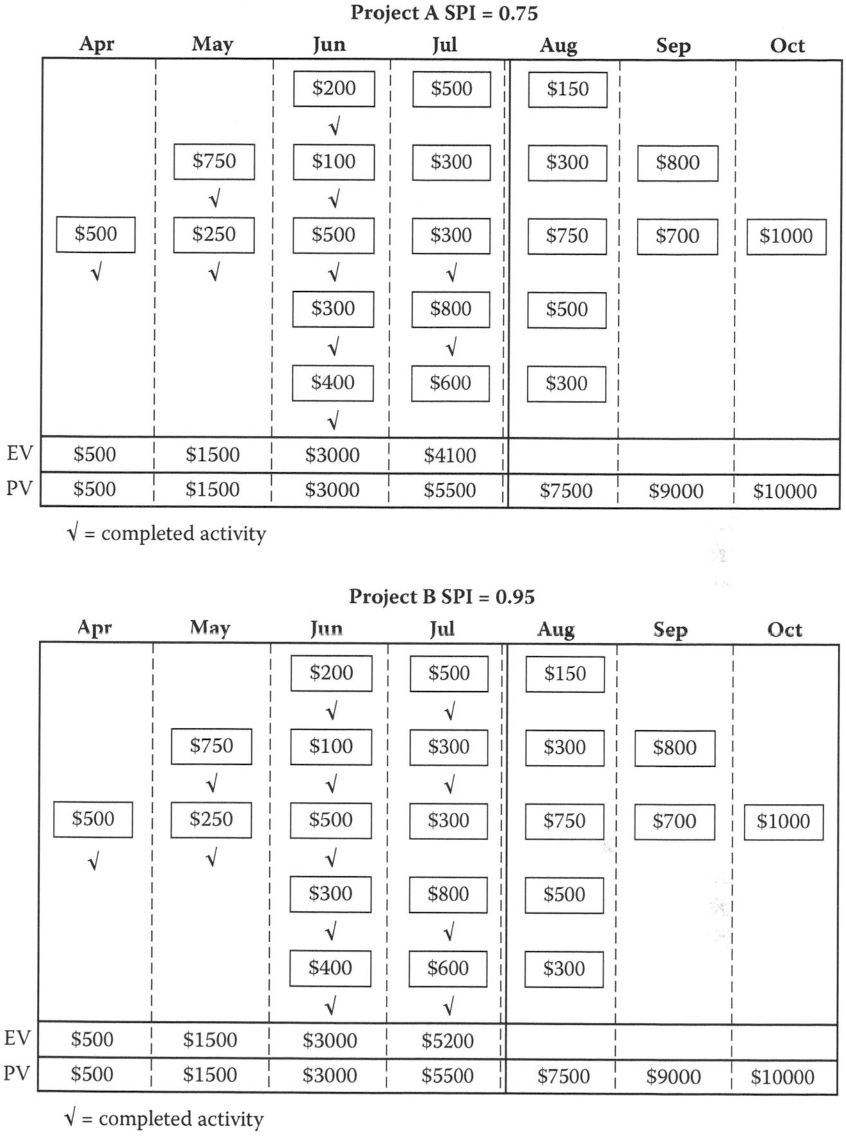

Project A SPI = 0.75

Apr	May	Jun	Jul	Aug	Sep	Oct
		$200 √	$500	$150		
	$750 √	$100 √	$300	$300	$800	
$500 √	$250 √	$500 √	$300 √	$750	$700	$1000
		$300 √	$800 √	$500		
		$400 √	$600	$300		
EV $500	$1500	$3000	$4100			
PV $500	$1500	$3000	$5500	$7500	$9000	$10000

√ = completed activity

Project B SPI = 0.95

Apr	May	Jun	Jul	Aug	Sep	Oct
		$200 √	$500 √	$150		
	$750 √	$100 √	$300 √	$300	$800	
$500 √	$250 √	$500 √	$300	$750	$700	$1000
		$300 √	$800 √	$500		
		$400 √	$600 √	$300		
EV $500	$1500	$3000	$5200			
PV $500	$1500	$3000	$5500	$7500	$9000	$10000

√ = completed activity

Figure 8.4 Two earned value tracking charts showing different SPIs.

But now let us look at Figure 8.5, where we have highlighted the critical path in each project, which just happens to run right down the middle. Now which project is closer to being on schedule? We can see that Project B is definitely behind schedule as its fourth critical path activity, worth $300 and scheduled to be finished in July, is not yet complete. Conversely, on

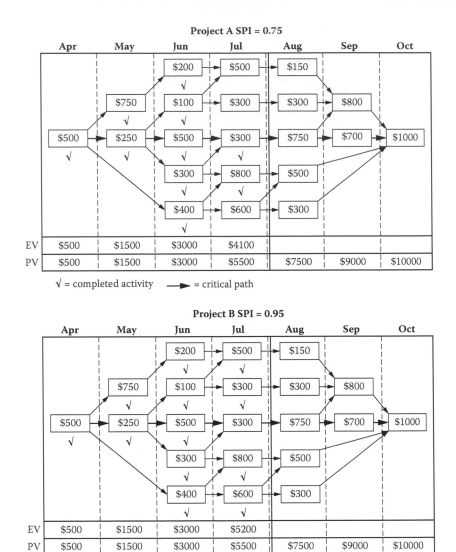

Figure 8.5 Same two earned value tracking charts but showing critical paths.

Project A, all the scheduled critical path activities have been completed. Although there are three incomplete activities worth a total of $1,400 that were scheduled to be completed in July, all of them are off the critical path and could have lots of float. It is therefore entirely possible that Project A, despite having a much lower SPI, is actually still headed for an on-time completion whereas Project B is definitely running late on its critical path.

Gaming the SPI

But there is another fundamental problem with the SPI as a schedule metric and that is the process of crediting a project with work completed ahead of schedule. As I said earlier, doing so makes perfect sense from a cost viewpoint: once an activity has been completed, we don't have to pay for it anymore. But giving credit for activities completed ahead of schedule makes absolutely zero sense from a schedule viewpoint unless the activity is on the critical path. An SPI of greater than 1.00 does not mean that the project is ahead of schedule and that the duration will be compressed unless it's the critical path activities that are achieving early completion.

The fact that SPI can be distorted by the total float of noncritical activities can combine with the practice of receiving credit for noncritical activities completed ahead of schedule to create a perfect storm of distortion. If a project manager is getting pressure for having a relatively poor SPI, one way to improve the metric is to complete activities that have big budgets, whether they're on the critical path or have lots of float. This means that not only is the metric distorted but it can actually cause counterproductive actions. Figure 8.6 shows us how bad things can happen.

At the end of June, Project C is on schedule on its critical path. However, two noncritical activities worth a total of $600 were scheduled for completion in June and are late. They may, of course, have plenty of float, so that if they are finished sometime in July they will not delay

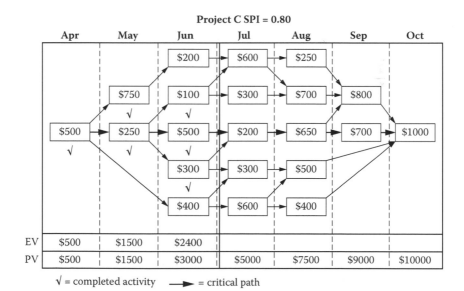

Figure 8.6 Earned value tracking chart showing slipping SPI.

the end of the project. However, the project manager issues a progress report and is thereupon beaten up (only metaphorically, of course) by the customer for having an SPI of 0.80. The customer really doesn't understand the concepts of critical path and float. All he knows is that the SPI should be 1.0 and it's not.

Now the project manager is concerned. Even though she's pretty sure that the project is still on schedule in terms of the critical path, she might never get to find out because her job is in jeopardy. She needs to do something about that SPI. And the next critical path activity has a budget of only $200, not enough to make a great difference to her SPI. Once she finishes the two activities that are currently late, she will need to throw lots more resources at their successor activities if they are not also to finish late and then her CPI will get into trouble. She only has so much time and resources available. What is she to do?

She does what many people in her situation have done before her: she decides to give her customer what he wants. If what matters to him is a good SPI, then she does what she needs to do to have a good SPI. The results of her decision are shown in Figure 8.7. She grabs the biggest budget items that she can get done during the month of July, not caring whether they are on the critical path or even if all of their logical predecessors have already been completed. She completes the two $600 activities that are due in July and then reaches into August and gets the $700 activity done even though one of its predecessors (with a puny $300 budget!) sits idle. Because completing that August activity in July still gives her

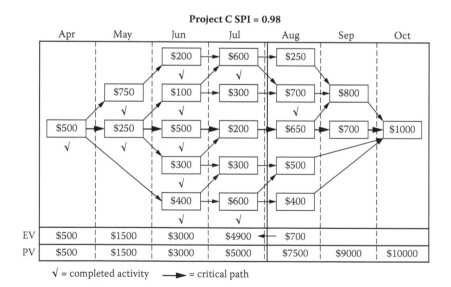

Figure 8.7 Improving the SPI by doing out-of-sequence work.

credit for it, the project's earned value rises to $4,900 and its SPI to a much healthier 0.98. And the customer is much happier, for a while.

Ultimately, the fact that the critical path is falling behind will become clear. The project will become later and later and, ultimately, more and more over budget. The question is only whether the project manager will have bought herself enough time to find another job before it's discovered that, despite the decent SPI, the project has fallen way behind schedule due to the slippage on the critical path.

Even more worrisome is that August activity that was performed out of sequence. Out-of-sequence work is often a project manager's nightmare. Engineers and other single contributors often take it upon themselves to decide that they really do not need to wait for the scheduled predecessor. The result is not only that progress reporting is fouled up and the schedule ceases to make much sense; it is also frequently the case that, despite the guilty engineer being unaware of it, there was a very good reason for that logical dependency and now work will have to be undone and redone to reflect the work scope of the predecessor, all the while causing delays and cost overruns.

Every project manager should, at the start of any project, make it clear that no one is allowed to do out-of-sequence work without first clearing it with the project manager. If an engineer hits upon an idea that may shorten the schedule by not waiting for a predecessor to finish, that's great. We want that engineer to step forward and, if the change is practical, to take credit for "thinking outside the box." But first, it must be approved and the schedule must be changed to reflect the new reality. Anyone who does work out of sequence without first getting the change approved should be quartered, drawn, and then hanged, and then told never to do it again.

Fixing the SPI

The two problems with the SPI can be fixed by two simple procedural changes:

1. Do not allow SPI credit for activities completed in a period earlier than scheduled unless (a) the activity was on the critical path as of the previous progress report and (b) all of its logical predecessors have been completed. This will encourage projects to speed up critical path activities but not through performing them out of sequence.
2. Use a separate PV baseline for schedule tracking than the one used for cost tracking. This should make sense under any circumstances; schedule is different from cost and is driven by different factors, primarily the critical path. The simplest adjustment would be to use what is sometimes called the ALAP schedule (as late as possible)

as the baseline PV for the schedule. In other words, place the PV activity or milestone dates for schedule tracking on those generated by the backward pass of the CPM algorithm. From Chapter 4, you may recall the network logic diagram in Figure 8.8 that shows the forward and backward pass dates. The backward pass dates are those at the bottom of the activity boxes, where there is no float and where anything that slips further will delay the end of the project. On the critical path, the dates will be unchanged as there is no float. But on noncritical activities, the baseline schedule for achieving them is their latest possible date without delaying the end of the project. As a result, the days on which the earned value of Activities B and D are scheduled to accrue would both be Day 18 (with float reduced to zero), the latest that those two activities can finish without delaying the end of the project. And if they are completed any earlier, no SPI credit will be given for them until Day 18.

Adopting these procedural changes would factor the distortions caused by float out of the SPI metric, as float would no longer be a part of PV for schedule. Currently, the traditional SPI can theoretically be 0.90 or even lower and the project still be on schedule if the 0.10 negative variance is all off the critical path and accounted for by float. But if the ALAP SPI for a project is 0.99, we can be sure that the project is running late because something must have slipped beyond its float to cause that 0.01 variance.

Of course, the only way now that the ALAP SPI could ever be greater than 1.00 is if, per Procedural Change #1 above, a critical path activity has been finished ahead of schedule. But isn't that what we want? Why would we want a metric to tell us that we are ahead of schedule if all

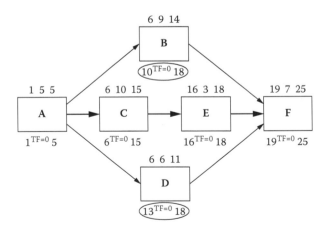

Figure 8.8 Schedule showing dates from the backward pass for an ALAP SPI baseline.

the variance comes from completing noncritical activities early (enough perhaps to offset the fact that our critical path may have slipped a bit and we're actually running late)?

Will such simple procedural changes ever be adopted? Most project management practitioners understand that the traditional SPI suffers from severe problems, and some have even tried to improve it.

Earned schedule tracking

The three tracking functions of earned value (PV, EV, and AC) are often referred to, due to their shapes, as the three S-curves.

- The PV curve, which is also the cost accrual curve, is the baseline schedule of how cost and earned value are expected to accumulate throughout the project.
- The EV curve is the schedule of how earned value has accumulated.
- The AC curve is the schedule of how costs have actually accumulated.

Figure 8.9 displays the three functions, with the PV at the current data date of the end of August at $2 million, the EV showing that $1.7 million of the planned work has been completed, and the AC showing that it has cost $2.1 million. Based on these metrics:

- The cost variance is $1.7 million – $2.1 million = – $0.4 million.
- The CPI is $1.7 million ÷ $2.1 million = $0.81.
- The schedule variance is $1.7 million – $2.0 million = −0.3 million.
- The SPI is $1.7 million ÷ $2.0 million equals 0.85.

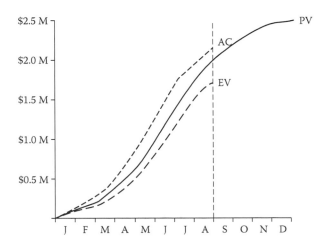

Figure 8.9 The PV, EV, and AC S-curves.

Notice the peculiarity of the schedule variance being measured in dollars. We are used to schedule being measured in time units, not monetary units. This irks a lot of people.

Another issue with earned value scheduling metrics is that at the end of the project the earned value is always equal to the planned value. As EV approaches 100%, the SPI becomes more and more stable and less and less useful. It's not at all clear what an SPI of 0.95 means on a project that was planned to take 10 months but that has already taken 20 months. To resolve these issues, a new earned value schedule tracking metric has gained popularity in recent years. It's called *earned schedule tracking*. The principle is shown in Figure 8.10.

In earned schedule tracking, the actual earned value at the data date is compared to the date at which that earned value was scheduled to have been achieved, that is, the date at which the PV function was at that number. In our example, the earned value of $1.7 million was scheduled to have been achieved at the end of July. In earned schedule terms, therefore, the project is one month, or 4.3 weeks, behind schedule. If the EV at the end of August had only been $1.5 million, that was the PV number at the end of June, and the project would be two months delayed in earned schedule terms.

If we are one month behind schedule after eight months of a 12-month project and we continue at this rate, our projection would be that the project will take 13.7 months. Using the SPI of 0.85, our duration at completion would be projected as 12 months ÷ 0.85 = 14.1 months.

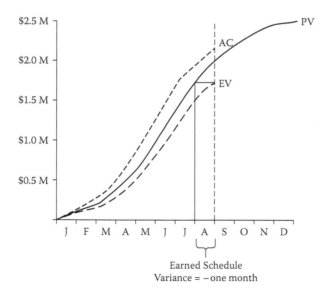

Figure 8.10 S-curves showing a negative earned schedule variance of one month.

Which is more accurate? Who knows? But we are definitely behind schedule … unless, again, the scheduled work that we have not completed is all off the critical path with lots of float.

Therefore earned schedule tracking solves the minor problems with earned value schedule tracking but not the major ones:

1. That earned value tracking is a cost-based metric
2. That cost and schedule are only tangentially related
3. That earned value does not recognize the most crucial (critical?) aspect of the project schedule, that is, the critical path
4. That earned value tracking processes continue to allow taking credit for schedule purposes of work performed ahead of schedule and even out-of-sequence

Tracking the DIPP through earned value

Whether we are on schedule or behind schedule, on budget or within budget, the most important questions that should be asked are

1. How is our project doing as an investment?
2. How much of the value that we expected to get out of it is still available?
3. How much more will we have to invest in order to get that value?
4. Is it possible that it will cost us more to finish this project than the value we will ever get out of it? Because if this is the case, we should terminate the project now.

It will be remembered from previous chapters that projects are performed for their expected project profit, which is their expected monetary value minus their cost. And that the metric by which we plan and track our project profit is the DIPP:

DIPP = ($EMV ± $[acceleration premium or delay cost]) ÷ $cost ETC

The assumption with most projects tends to be that, if it is completed on the target date (usually and unfortunately codified as a deadline), that the expected value will be exactly what it was when planning began months or years before. This obviously is an unwarranted assumption; all kinds of external factors in terms of international situations, economic conditions, competitive markets, and other variables can have a dramatic impact on both the expected value and the value/cost of time on which the acceleration premium and delay cost modifiers of value are based. Changes in these should trigger changes in the project plan. However, these values are often never tracked or updated during project performance, unlike cost ETC, which is regularly updated through earned value metrics.

We know that the project is scheduled to be finished at the end of December for a total cost of $2.5 million. We also know that at the end of August, the PV was $2 million, meaning that the planned cost estimate-to-complete at the end of August was $0.5 million.

Now let us stipulate three items about the project:

1. If completed December 31, the project's EMV will be $5 million.
2. The acceleration premium for every full week early is $100,000.
3. The delay cost for every week or part thereof late is $400,000.

With a starting EMV of $5 million, a planned completion date of December 31, and a budget of $2.5 million, the project's starting DIPP is 2.0. But if the EMV remains constant and there is no acceleration or delay, then at the end of August the only thing that will have changed for sure is the planned cost ETC, which is predicted to be $0.5 million. In that case, the planned DIPP at the end of August will be $5 million ÷ 0.5 million, or 10.0.

But from our earned value reports, we know that the project is running both over budget and behind schedule. Based on the CPI of 0.81, the project's cost estimate-at-completion is $2.5 million ÷ 0.81 = $3,086,420 for a cost overrun of $586,420. Because our sunk cost (i.e., AC) at the end of August is $2.1 million, our cost ETC = $3,086,420 − $2,100,000 = $986,420.

For our purposes, we assume that the SPI on this project is an accurate reflection of schedule progress (an assumption that would be much more justified if the project were using an ALAP SPI). If we work at the current 0.85 SPI throughout our project's 52-week schedule, the duration estimate-at-completion should be 52 weeks ÷ 0.85 = 61.2 weeks. This will mean a delay of 10 weeks or parts thereof at a cost of $400,000 per week, or a total delay cost of $4 million.

It is immediately very easy to see that this project is headed for huge losses. With the delay costs, the project EMV has fallen to $5 million − $4 million = $1 million. Meanwhile the CPI is forecasting a cost EAC of $3,086,420, for a loss of $2,086,420. Faced with a project that is so challenged in terms of both schedule and cost, many organizations would decide to terminate it immediately. But the decision is actually not such a simple one. The reason is that $2,100,000 of that $3,086,420 is sunk costs, irretrievable whether we cancel the project now or keep on going. From an investment point of view we should only decide to cancel the project if we believe that the value we will get out of it will be less than what it will cost to finish it. And analysis of the detailed data turns out to be borderline:

Actual DIPP = ($5,000,000 − $4,000,000) ÷ ($3,086,420 − $2,100,000) = 1.01

DIPP Progress Index (DPI) = Actual DIPP ÷ Planned DIPP = 1.01 ÷ 10.00
= 0.10

This is telling us that, whereas when we planned the project we had expected that every dollar we invested after the end of August would generate a return of $10, that has now fallen to one dollar and one penny. If this is in fact the reality, then it's certainly possible that we should terminate the project. But first there is a need for even more detailed cost–benefit analysis.

The foundation for such analysis goes back to the very first article I published on the DIPP, "When the DIPP Dips" in the September–October 1992 issue of *Project Management Journal* and which was republished as a chapter in the Project Management Institute's 1999 book, *Essentials of Project Control*, edited by Pinto and Trailer. When a project's DIPP nears the 1.00 threshold that means it's no longer profitable to continue, there are several factors which might make project termination the better choice. These include opportunity costs (i.e., more profitable ways of investing that same money) and salvage value (value that has already been created and would remain even if we terminate the project before completion). Any project being considered for termination should be subjected to rigorous analysis, including those costs often associated with terminating a project before completion, which can include angry customers, bad public relations, empty buildings, and increased unemployment insurance.

But any investment that suggests, *prima facie*, a return of just 1% at the end of six months is probably a candidate for amputation, at least if we assume that the SPI projections are accurate. This only serves to emphasize that, if the value/cost of time on a project is substantial, it is crucial that we use schedule metrics which are not distorted.

How to use the DIPP to redeem projects

Earned value tracking allows us to measure how we're doing in terms of cost and, bearing in mind the caveats about the SPI's accuracy, in terms of schedule. Those metrics can then be plugged into the DIPP formula to generate the cost ETC, to produce the actual DIPP and the DPI. We can then engage in analysis and decision making based on the project's investment value.

One of the nice things about projects as investments is that, if we manage them as such, we can actually exercise a degree of control that is often not present in other kinds of investments. The purchase of a US Treasury security or a corporate stock or a parcel of waterfront land is almost completely driven by factors that are beyond our control. But with a project, provided we know what we are doing, we can often make decisions to improve the value of the investment.

Just as a project should not be pigeonholed in the binary fashion as a "success" or "failure," so too decisions about a project should not be reduced to "terminate" or "continue according to the current plan." If analysis of our current predicament discloses the reasons, then we very likely can use the techniques described earlier in the book to reshape the project and redeem at least some of our investment's value.

It's easy to see why our project has lost so much expected value: the cost of the 10-week delay has reduced the EMV from $5 million to $1 million. The cause of the delay might reveal ways to improve the outlook. But under any circumstances, we likely need to change our plan for the remaining months of the project.

The delay, if the SPI and its projections are accurate, means that the critical path is pushing out 10 weeks longer than planned. There are activities on the remaining critical path that have drag and drag cost, and the drag cost for up to 10 weeks of drag is now $400,000 per week. Surely we can find ways of adding $5,000 or $10,000, or $50,000, or even $200,000 that will allow us to reduce drag on some of the remaining activities by a week or more. Every such opportunity that we identify provides the chance to improve our EMV. And it wouldn't take many such amendments to change our decision data from its current heads-or-tails balance to one where completing the project is clearly the right decision.

If by spending an extra $400,000 we can be sure that we will reduce the delay from 10 weeks to just 7 weeks, the reduction in delay costs will be worth $1,200,000 and the EMV will rise from $1 million to $2.2 million. The cost ETC will also rise, but only by $400,000, from $986,420 to $1,386,420. Now the project will have an expected value of $5,000,000 minus delay costs of 7 weeks at $400,000 each, or $2,200,000. With the added funding, the cost EAC will now be $3,086,420 plus $400,000, or $3,486,420. We are still projected to lose $1,286,420.

But even though the project will lose money, it will lose less money than if we either terminate it now (which would result in a loss of the $2.1 million of sunk costs) or continue according to the current plan (which is forecast to result in a loss equal to the current EMV minus delay cost ($1,000,000) minus the current cost EAC ($3,086,420), or $2,086,420. Would we rather lose this amount, or $800,000 less?

With this change the actual DIPP would now be $2,200,000 ÷ $1,386,420 = 1.59. A 59% return in six months or less does not at all seem like a bad way to invest that $1,386,420. Of course, if our organization is cash-strapped, we may not be able to do this: if you don't have the money, you can't spend it. But if we do have the money, and there isn't a better way of investing that additional $400,000, then the additional investment is the best alternative.

Summary points

1. Earned value tracking is a valuable technique for cost control on large projects. However, although its use has spread considerably in the past two decades, full appreciation of both its uses and its shortcomings is rare.

2. Earned value is not about value; it's about cost!
3. Earned value schedule tracking metrics disregard the most important factor that determines a project's duration: the critical path. This can be corrected by developing a second earned value baseline for schedule tracking, distinct from cost tracking. This second baseline should schedule earned value to accrue on the as-late-as-possible dates generated by the backward pass of the critical path algorithm, where there is no float and everything is critical.
4. Because schedule float can distort the earned value schedule tracking metrics, there is the danger of the metrics being "gamed" by prioritizing the performance of noncritical activities that have lots of earned value over activities that are on the critical path. This often winds up making a late project even later.
5. Allowing earned value credit for activities performed ahead of schedule can also lead to the project team doing work out of sequence. To correct this, the earned value of activities achieved ahead of schedule should be credited in the scheduling metrics only if it is an activity with zero or negative float.

chapter nine

Advanced earned value

"How can I tell what earned value is really saying?"

Chapter 8 summarizes the workings and the purpose of earned value and earned value tracking: what it is, what it isn't, how to fix some of its faults and distortions, and how to use it. If you previously had only a limited sense of the fundamental workings of earned value, Chapter 8 was intended to inform you about the basic techniques. If you feel comfortable with that information, you probably have a better comprehension of earned value, both its appropriate uses and its shortcomings, than most of its practitioners.

However, a little more needs to be said about ways in which earned value metrics can be extended to do an even better job of managing projects. The standard data can be sliced and diced in different ways and used in additional formulas to give more information about the progress of a project.

Earned value based on milestones

One of the criticisms of earned value tracking is that it often relies on estimates of percent complete. This means that if an activity is ongoing at the time of a progress report (i.e., it has been started but not yet been completed), then earned value can be assigned to the work done on that activity based on an estimate of what percentage of it has been completed. So if an activity with a budget of $10,000 is reported as 40% complete, the project will be credited with $4,000 of earned value. Most software packages support this.

The problem with this is the subjective nature of percent complete estimates. Indeed, no team member should ever be asked to estimate the percentage completion of his or her activity. If he reports it as "60% complete," here is what he could justifiably mean, without any attempt to deceive:

- It has used up 60% of its original duration estimate.
- It has used up 60% of the time he now realizes it will really take.
- It has used up 60% of its original budget.
- It has used up 60% of the money he now realizes it will really take.
- He lifted the lid on the stewpot, stuck his head in, sniffed, and decided it was 60% cooked.

Now throw in a dash of pressure because the customer is getting antsy, and such an estimate becomes completely worthless. No project manager therefore should ever ask for a percent complete estimate; instead, she should ask:

1. How much longer till it's finished?
2. How much more will it cost till it's finished?

Then the project manager can determine what percent complete she thinks the activity is. But even a project manager's estimate can be tainted by subjectivity.

Almost all US Defense Department programs that use earned value track it on the basis of activity-driven *milestones*. A milestone is an instant in time, like an activity with a duration of zero. Therefore there is never any need to estimate a milestone's percent complete; it has either been achieved and is 100% complete or it hasn't been and is 0% complete. An activity-driven milestone is the finish or start of an activity or work package. And the earned value weight of an activity-driven milestone is based on the budget of the activity that is driving it. There are three basic types of weighting systems:

1. 0–100 weighting, where no earned value is credited until the activity is completed at which time it gets the earned value of 100% of the driving activity's budget.
2. 50–50 weighting, where half an activity's earned value is credited as soon as the activity starts and the other half is credited when the activity finishes.
3. 25–75 weighting, where 25% of an activity's earned value is credited at the start and the other 75% when it finishes.

Figure 9.1 shows a diagram of an earned value baseline chart with a milestone displayed as a triangle at the scheduled finish of every activity. This is a chart for a 0–100 weighting system, with the PV calculated for when the activity is scheduled to finish. Indeed, there is no need even to show the activity; we can simply show the milestones on their schedule dates, as displayed in Figure 9.2.

Figure 9.3 shows which milestones have been achieved as of the end of the July data date. A total of $500 worth of milestones that were scheduled have not yet been achieved, making the project's earned value $3,500 compared to its planned value of $4,000. This represents a schedule variance of –$500 and a schedule performance index of $3,500 ÷ $4,000 = 0.88.

In general, contractors do not like performing projects that are being tracked on the 0–100 basis. The reason is that they have to spend the money on the work in the first 99% of each activity, but only get the earned value credit when they do that last one percent. As a result, the cost metrics

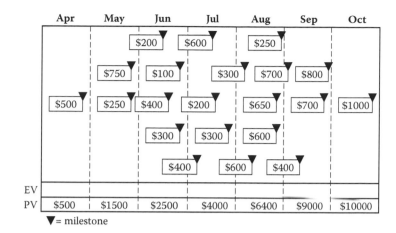

Figure 9.1 A finish milestone chart, weighted at 100% of the driving activity's budget.

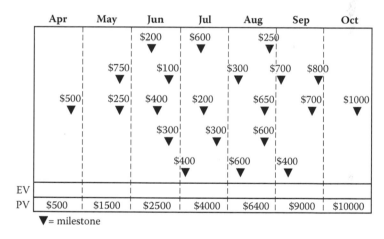

Figure 9.2 A 0–100 milestone chart showing only the milestones, not the activities.

tend to show them over budget. If we look at Figure 9.3, the project has almost certainly spent money on the two incomplete activities scheduled for completion in July, and very likely has spent money on the two activities scheduled for completion early in August. Yet it receives no earned value credit whatever for this work.

Contractors therefore much prefer the 50–50 weighting system, where there are two milestones for each activity, representing the start and finish. Half the budget for the activity is given as earned value credit as soon as we start the activity and the other half when we finish it. Figure 9.4 shows a milestone chart with the 50–50 weighting system.

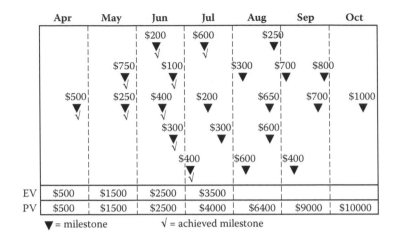

Figure 9.3 A 0–100 milestone chart showing earned value.

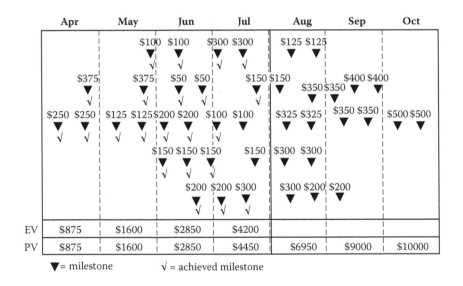

Figure 9.4 Earned value tracking chart using the 50–50 milestone weighting system.

In Figure 9.4 we can see that using the 50–50 weighting system has accelerated the PV totals, but has also allowed earned value to be achieved earlier. At the end of July, our PV is $4,450. But our EV has also risen to $4,200 as a result of having achieved the start milestones. Our SPI is all the way up to 0.94 and, with credit for more earned value, our CPI is also undoubtedly higher.

So why isn't the 50–50 weighting system always used? The answer is because it too can distort reality. It is very easy to game the 50–50 system simply by starting activities and doing minimal work on them. Once we've started them we get credit for half of the driving activity's budget worth of earned value, even though we may only have spent $10 on the activity. Of course, this is fraud, and a project manager for a US government contractor who engaged in such shenanigans would be risking prison. Nevertheless, such behavior can be difficult to identify.

The 25–75 weighting system is like the 50–50 system except that the start milestone is only worth 25% of the activity's budget. This makes sense, because finishing an activity is much more valuable than starting it; technical problems tend to arise between those two milestones. It also provides less incentive for the sort of gaming we identified above. It is probably the best method of milestone weighting, but it is far less common than the 50–50 method.

Tracking CPI based on labor

On many large projects, and particularly US Department of Defense programs, earned value lumps all costs within the budget together: materials, equipment, labor, travel, and so on. This certainly helps to create a sense of how the project is doing in terms of its usage of all resources that are included in the budget. However, this sometimes also can lead to distortions. Different resources have different characteristics. Some are much more expensive than others, and some are much dodgier and riskier in their technical challenges than others. Both of these factors can make any metric that lumps them all together undependable.

Activities that are labor-intensive tend to be the most volatile both in duration and resource usage. Yet because materials and equipment are often much more expensive than labor, the budgets for activities that are heavy in these types of resources can greatly outweigh those risky activities that require labor. This leads to what is sometimes called the purchased items distortion, where activities that seem to be lower risk because of their small budgets devoted to labor often far outweigh, in terms of equipment budgets and earned value, the labor-intensive activities with smaller budgets but greater volatility and often higher risk.

Another drawback of lumping all the resources together into a single budget for earned value purposes is the fact that it requires complex financial systems to track all the different types of cost and to provide timely reports. Large engineering companies usually have sufficiently robust systems, but many smaller organizations that could derive benefit from earned value tracking on their larger projects do not have such functionality.

All of the above issues can be resolved by using what is sometimes referred to as CPI-Labor, which does the earned value weighting based only on labor, either labor hours or labor budget. For large and labor-intensive projects, such as in corporate IT departments, earned value analysis and tracking can be enabled simply by budgeting activities in labor hours and then requiring employees to report their time against each activity. But even organizations with complex financial systems can generate valuable analysis by separating out the labor component of the budget and using two earned value baselines: one based on total budget and one based on labor hours. This approach is now standard for US Defense Department contractors. Earned value metrics based on labor hours not only removes the distorting aspects of material and equipment purchases but also provides additional insight.

In Chapter 3, I mentioned retired US Navy Lieutenant Commander Joe Sopko, who learned about the DIPP and critical path drag from a PMI seminar I taught in Baltimore in 2002. Wherever Joe has worked since, he has deployed them. But over the years, I too have learned many things from Joe. One valuable nugget is a point that he makes about budget-based earned value metrics for a project team with exempt employees who are not paid for overtime. Figure 9.5 shows the plan and actuals for the first four weeks of a project.

As the table shows, everything seems to be going according to plan. Item 5 shows that we have gotten 100% of the planned work accomplished, so the project is on schedule. The work was planned for 40 hours per week, or a total of 160 hours. The planned cost of the work was $50 an hour,

	Week 1	Week 2	Week 3	Week 4
1. Planned Work Hours	40 Hrs.	40 Hrs.	40 Hrs.	40 Hrs.
2. Planned Cum. Hours (Labor PV)	40 Hrs.	80 Hrs.	120 Hrs.	160 Hrs.
3. Planned Wkly Cost (@ $50/Hr)	$2,000	$2,000	$2,000	$2,000
4. Planned Cum. Cost (PV)	$2,000	$4,000	$6,000	$8,000
5. Planned Work Accomplished	100%	100%	100%	100%
6. Total Budget Earned Value (EV)	$2,000	$4,000	$6,000	$8,000
7. Actual Wkly Cost	$2,000	$2,000	$2,000	$2,000
8. Actual Cum. Cost (Tot. Budget AC)	$2,000	$4,000	$6,000	$8,000
9. Total Budget SPI (6 ÷ 4)	1.00	1.00	1.00	1.00
10. Total Budget CPI (6 ÷ 8)	1.00	1.00	1.00	1.00

Figure 9.5 Planned and actual labor hours, cost, and EV metrics for a sample project.

or $8,000 for the first four weeks of work. Because we have completed 100% of the planned work our earned value thus far, based on budget, is $8,000 and the cumulative actual cost (Item 8) is $8,000. These metrics give us both an SPI and a CPI of 1.00.

What these data are not showing is that the project has stayed on schedule by its fingernails. The employee doing the work has been per-forming heroically, willing to work 20 hours of overtime each week, or a total of 240 hours during the four weeks. But how long will he be able to keep it up? Numerous studies over the years have shown that even when employees are willing to work huge numbers of hours week after week, sooner or later their productivity starts to suffer. Even if they continue to work the extra hours, their work performance will decline and the project will fall behind schedule.

This leads to a well-known phenomenon in earned value track-ing: projects often are able to stay on schedule only through additional resource usage. This means that a declining CPI, caused by the extra resource usage, is often a leading indicator that the SPI is also going to decline and the project is going to fall behind schedule.

In our project above, if this employee who is working so hard were nonexempt and being paid time-and-a-half for the overtime hours, we would be able to see that the project was actually way over budget and the CPI far below 1.0. But if the employee is exempt, then he is being paid his weekly salary whether he works 40 hours or 60 hours, so that the AC is not affected by the overtime hours. Thus the organization does not have that declining CPI to act as a canary in the coal mine and the sponsor/customer assumes that things are going well, even though the chutney is about to hit the ventilation.

There is one way to bring visibility to the rising problem: creating a base-line based on labor hours and then tracking a CPI based on only the actual labor resources. If we had done that on the project above, what we would see is the chart shown in Figure 9.6. As we can see, although the fact that the exempt employee's dollar cost is fixed for every week of work, our project is actually going over budget every week in terms of the planned work hours. Our labor CPI during the first four weeks has been 0.67, meaning that we are staying on schedule only by using 50% more labor than planned. That should be a signal to the project manager, the sponsor, and the organization that the budget is insufficient to keep this project on schedule.

Finally, during Week 5 our hero is no longer able to provide 60 hours at the previous productivity level to our project. Perhaps his functional manager says: "Hey, Henry, if you're going to work 20 hours extra every week, fine. But not on that project. I only promised you to that project for 40 hours a week. I want you to start working the extra 20 hours on this other project that's just starting up." Or perhaps Henry collapses from exhaus-tion. Perhaps his concentration level falters. Perhaps it's a combination

	Week 1	Week 2	Week 3	Week 4
1. Planned Work Hours	40 Hrs.	40 Hrs.	40 Hrs.	40 Hrs.
2. Planned Cum. Hours (Labor PV)	40 Hrs.	80 Hrs.	120 Hrs.	160 Hrs.
3. Planned Wkly Cost (@ $50/Hr)	$2,000	$2,000	$2,000	$2,000
4. Planned Cum. Cost (PV)	$2,000	$4,000	$6,000	$8,000
5. Planned Work Accomplished	100%	100%	100%	100%
6. Total Budget Earned Value (EV)	$2,000	$4,000	$6,000	$8,000
7. Actual Wkly Cost	$2,000	$2,000	$2,000	$2,000
8. Actual Cum. Cost (Tot. Budget AC)	$2,000	$4,000	$6,000	$8,000
9. Total Budget SPI (6 ÷ 4)	1.00	1.00	1.00	1.00
10. Total Budget CPI (6 ÷ 8)	1.00	1.00	1.00	1.00
11. Actual Wkly Hours	**60 Hrs.**	**60 Hrs.**	**60 Hrs.**	**60 Hrs.**
12. Cum. EV by Labor Hours (Labor EV)	40 Hrs.	80 Hrs.	120 Hrs.	160 Hrs.
13. Cum. Wkly Hours (Labor AC)	**60 Hrs.**	**120 Hrs.**	**180 Hrs.**	**240 Hrs.**
14. Labor CPI (12 ÷ 13)	**0.67**	**0.67**	**0.67**	**0.67**

Figure 9.6 The labor CPI metric shows that labor has been greater than planned.

	Week 1	Week 2	Week 3	Week 4	Week 5	Week 6
1. Planned Work Hours	40 Hrs.	40 Hrs.	40 Hrs.	40 Hrs.	40 Hrs.	40 Hrs.
2. Planned Cum. Hours (Labor PV)	40 Hrs.	80 Hrs.	120 Hrs.	160 Hrs.	200 Hrs.	240 Hrs.
3. Planned Wkly Cost (@ $50/Hr)	$2,000	$2,000	$2,000	$2,000	$2,000	$2,000
4. Planned Cum. Cost (PV)	$2,000	$4,000	$6,000	$8,000	$10,000	$12,000
5. Planned Work Accomplished	100%	100%	100%	100%	**50%**	**25%**
6. Total Budget Earned Value (EV)	$2,000	$4,000	$6,000	$8,000	$9,000	$9,500
7. Actual Wkly Cost	$2,000	$2,000	$2,000	$2,000	$2,000	$2,000
8. Actual Cum. Cost (Tot. Budget AC)	$2,000	$4,000	$6,000	$8,000	$10,000	$12,000
9. Total Budget SPI (6 ÷ 4)	1.00	1.00	1.00	1.00	**0.90**	**0.79**
10. Total Budget CPI (6 ÷ 8)	1.00	1.00	1.00	1.00	**0.90**	**0.79**
11. Actual Wkly Hours	**60 Hrs.**	**60 Hrs.**	**60 Hrs.**	**60 Hrs.**	40 Hrs.	40 Hrs.
12. Cum. EV by Labor Hours (Labor EV)	40 Hrs.	80 Hrs.	120 Hrs.	160 Hrs.	**180 Hrs.**	**190 Hrs.**
13. Cum. Wkly Hours (Labor AC)	**60 Hrs.**	**120 Hrs.**	**180 Hrs.**	**240 Hrs.**	**280 Hrs.**	**320 Hrs.**
14. Labor CPI (12 ÷ 13)	**0.67**	**0.67**	**0.67**	**0.67**	**0.64**	**0.59**

Figure 9.7 Total budget CPI and SPI start to slip as predicted by CPI labor.

of two of the above, including the difficulty of going from immersion in one project to trying to concentrate on two. Under any circumstances, in Week 5 Henry only works on our project for 40 hours and is no longer able to accomplish all the scheduled work. As Figure 9.7 shows, the schedule starts to slip, and continues to do so in Week 6.

This situation was always a disaster waiting to happen. Unfortunately, because we were not able to see that our original estimates for resource usage were inadequate, we have been caught completely by surprise. Now we have fallen behind schedule and we might never be able to catch up. And the delay costs on this project could be huge. Had we tracked work hours and labor CPI, we would have been able to see the problem earlier and we might have been able to take corrective action, such as assigning an additional resource to help Henry.

To complete performance index (TCPI)

When a project manager is told that the CPI is below par and the project is trending to be well over budget, she often becomes quite defensive and asserts that, despite the current data, things are not as bad as they look. "Don't worry," she says. "We'll figure it out and finish on budget." There are many studies showing that, once a project passes the 20% point in earned value, if its CPI is below .90 it is almost impossible to recover. But it is nice to have numbers that clearly demonstrate the improbability of recovery.

As an example, let us take a project with a $10 million budget. Let us assume that our EV is $4 million, meaning that we have achieved 40% of the earned value. And let us assume that we have spent $5.5 million thus far. Our CPI is therefore $4 million ÷ $5.5 million = 0.73. The question is: what will our CPI have to be for the rest of the project if we are to finish on budget?

The formula we use is called the to complete performance index (TCPI). Essentially, this involves comparing how much we have been spending for the portion of the earned value we have achieved thus far and how much money is left in the budget to complete the rest of the work. The formula is

TCPI = (Budget-at-Completion − EV) ÷ (Budget-at-Completion − AC)

= ($10 million − $4 million) ÷ ($10 million − $5.5 million)

= $6 million ÷ $4.5 million

= 1.33

For the first 40% of the project, we have worked at a CPI of 0.73. To finish on budget, we need to perform the remaining 60% at a CPI of 1.33. How likely do you think that is?

Critical ratio cost index

We know that the longer a project goes on, the greater the cost is likely to be. As we discussed earlier, the reason is primarily due to "marching army" costs: project support activities, overhead, and other indirect costs that will continue until the project ends. This may be one reason why cost estimates based on the CPI often turn out to be optimistic. If the project is both over budget and behind schedule, the overspending will go on for longer than planned.

One index that has been shown to give a more accurate prediction of final cost than the simple CPI is called the critical ratio cost index (CRCI). It is the product of both the CPI and the SPI: the CPI multiplied by the SPI. In the $10 million project above with the EV of $4 million and the AC of $5.5 million, let us assume that the PV is $5 million. Using these numbers, we can compute:

$$CPI = \$4 \text{ million} \div \$5.5 \text{ million} = 0.73$$

$$SPI = \$4 \text{ million} \div \$5 \text{ million} = 0.80$$

$$\text{Critical Ratio Cost Index} = 0.73 * 0.80 = 0.58$$

We can usually get a more accurate cost estimate-at-completion by dividing the budget-at-completion by the CRCI:

$$\text{Cost EAC} = \$10 \text{ million} \div 0.58 = \$1,724,138$$

Earned value based on milestones

Finally, there is one other significant shortcoming of the CPI. There are two very different genres of problems that a project can run into, and each has completely different implications for cost. The two types of problems are

1. One-time hits that have a sharp impact and are then resolved
2. Continuing trends that cause a project to bleed money throughout execution

The problem is that the project CPI assumes that all causes of over-spending are of the second type. It applies the CPI for the early part of the project throughout the rest of the project in a "straight line" function. This may not always make sense; there are often technical problems that crop up during the design phase of a project that require unexpected time and money to solve. But once they are solved, why would we "charge" the rest of the project to fix a problem that has already been resolved?

One way that organizations with mature earned value processes overcome this issue is by planning and tracking separate earned value metrics for each of the different functional areas that will be performing work on the project. They can then see how each department has been performing against its budget, isolate the impact of overspending in those areas, and conduct trend analysis based on functional department performance. Figure 9.8 shows how this might look for a project within an engineering organization. The project has a budget of $10 million. Activities/milestones that were budgeted for a total of $6 million have thus far been accomplished, giving us that much in earned value. Accomplishing that work has actually cost us $7 million. The project CPI is therefore $6 million divided by $7 million = 0.86.

However, as we can see, the different types of work activities that were assigned to the different functional departments have performed to budget with varying degrees of success. When we look at the CPI for each department, we can see a range from a 0.75 CPI for Systems Engineering to 1.00 for Project Management. We can also see that the percentage of assigned work that each department has performed thus far varies: the Systems Engineering Department has completed almost all its work on this project whereas the Test Engineering Department has completed only half. This means that although the Systems Engineering Department may continue to overspend its budget for the rest of the project (due perhaps to a specific design problem), it should not have that much of an impact on the project the rest of the way. Figure 9.9 shows the cost estimate-at-completion for each department and then sums them to get a bottoms-up cost EAC for the whole project.

As Figure 9.9 shows, the bottoms-up department-by-department CPI forecasts a cost EAC of $242,008 less than is calculated by using a single CPI across all the project work. The reason is that the overspending has been identified with the work of specific departments and those departments' activities on this project are almost complete.

Department	1. Budget	2. Earned Value	3. Actual cost	4. CPI (2 ÷ 3)
A. Systems Engineering	$2,000,000	$1,800,000	$2,400,000	0.75
B. Mechanical Engineering	$1,500,000	$900,000	$1,000,000	0.90
C. Electrical Engineering	$2,500,000	$1,400,000	$1,500,000	0.93
D. Software Engineering	$2,500,000	$1,200,000	$1,300,000	0.93
E. Test Engineering	$1,000,000	$500,000	$600,000	0.83
F. Project Management	$500,000	$200,000	$200,000	1.00
H. Total at Project Level	$10,000,000	$6,000,000	$7,000,000	0.86

Figure 9.8 EV chart showing project CPI broken down by functional departments.

Department	1. Budget	2. Earned Value	3. Actual cost	4. CPI (2 ÷ 3)	5. Cost EAC (1 ÷ 4)	6. Cost Overrun (5 – 1)	7. Cost EAC (5 – 3)
A. Systems Engineering	$2,000,000	$1,800,000	$2,400,000	0.75	$2,666,667	$666,667	$266,667
B. Mechanical Engineering	$1,500,000	$900,000	$1,000,000	0.90	$1,666,667	$166,667	$666,667
C. Electrical Engineering	$2,500,000	$1,400,000	$1,500,000	0.93	$2,659,574	$159,574	$1,159,574
D. Software Engineering	$2,500,000	$1,200,000	$1,300,000	0.93	$2,688,172	$188,172	$1,388,172
E. Test Engineering	$1,000,000	$500,000	$600,000	0.83	$1,204,819	$204,819	$604,819
F. Project Management	$500,000	$200,000	$200,000	1.00	$500,000	$0	$300,000
G. Total Summed from Depts. (A-F)	$10,000,000	$6,000,000	$7,000,000		$11,385,899	$1,385,899	$4,385,899
H. Total at Project Level	$10,000,000	$6,000,000	$7,000,000	0.86	$11,627,907	$1,627,907	$4,627,907
Difference (H minus G)					$242,008	$242,008	$242,008

Figure 9.9 EV chart with difference in cost EAC when based on CPI for each department.

Combining earned value metrics with the DIPP

Finally, let us assume that at the start of this project we had estimated an expected monetary value of $14.5 million if completed in 50 weeks, for a starting DIPP based on the target date of 1.45 ($14.5 million divided by $10 million). Let us also assume an estimated delay cost of $1 million per week, and that our schedule analysis (based on both SPI and critical path slippage) indicates that we will finish 10 weeks late, in 60 weeks. The forecast delay cost decreases the project's EMV from $14.5 million to $4.5 million.

We would, of course, search for ways to avoid the schedule delays, perhaps by adding resources (and increasing our cost ETC) or by cutting scope (which would decrease the EMV). In fact, we should have been doing this all along; we should be doing this continuously on every project. But what if we cannot find a way to pull in the schedule? What if our fate of being 10 weeks late is written in stone? Should we terminate the project or should we pay to complete it? As you may already have noticed, it could all depend on which cost ETC we use: the one for the project as a whole or the one calculated department by department.

- If we use the former, the cost ETC is $4,627,907. If the project's expected monetary value is $4,500,000, we would be $127,907 better off if we cancel the project now.
- If we use the latter department-based calculation, the cost ETC is $4,385,899 and we would be $114,101 better off if we complete the project.

So which calculation is correct? The department-based CPI gives us more information and thus we may be better off depending on it. But the numbers here are close enough that our company is not likely to be driven out of business on the basis of this one decision.

Unfortunately, decisions which do not recognize that every project is an investment and that we must make decisions based on investment value result in huge losses in the business world every day. And the sum of those decisions sometimes does drive companies out of business, and sometimes causes the loss of many human lives.

Summary points

1. Tracking work accomplished and earned value on the basis of estimates of "activity percent complete" suffers from imprecision, subjectivity, and wishful thinking. A better way of tracking earned value is on the basis of activity-driven milestones, where the milestone represents the instant an activity starts or finishes.
2. The 0–100 milestone weighting method gives credit for all of an activity's earned value only when the activity is completed. This means no earned value credit for an activity that is 90% finished and has spent 90% of its budget, making both the activity and the project appear to be over budget. This method of milestone weighting is not popular with contractors.
3. The 50–50 milestone weighting method gives half the credit for an activity's earned value as soon as it starts and the rest when it finishes. This method is subject to "gaming" by starting lots of activities but doing almost no work on them. Overall, a 25–75 weighting system is probably better in that it weights a finish as three times more valuable than a start.
4. Isolating and tracking the usage of labor hours with a separate "labor CPI" has great value, especially on a project that uses exempt employees whose overtime hours may otherwise not be counted. When the labor CPI deteriorates, it often means that the project is only staying on schedule by using far more labor hours than planned, and this is often an omen that schedule slippage is about to occur.
5. The to complete performance index (TCPI) computes what this CPI would have to be for the rest of the project in order for it to finish on budget. This index is often used to refute a project team's claims of "Don't worry. We'll make it up the rest of the way and finish on budget."
6. The critical ratio cost index (TRCI) takes into account that if a project is both over budget and behind schedule, the overspending will continue for longer than planned. The TRCI is the product of the CPI

and SPI and is used to make cost EAC forecasts: cost EAC equals budget-at-completion ÷ (CPI * SPI).

7. A project's CPI may often be affected by a "one-time hit" causing overspending in one technical area, such as design. To avoid such an isolated occurrence "infecting" the CPI for the entire project, it is often more accurate to predict the project's cost EAC by collecting separate CPIs and estimates-at-completion for each functional department.

chapter ten

Conclusion

"What should all this mean for me?"

Over the course of the preceding chapters, I have laid out a new approach to planning and managing projects. That approach starts with a new definition: every project is an investment. If this stipulation is accepted, it follows that projects should be managed in a way that is similar to the way all other investments are managed, and that decisions should be made on the basis of the guiding metric for all investments: the value of the project's return above cost.

Most people who work in business and government will agree that the performance of projects has significant opportunity for improvement. The discipline of project management has developed a large toolbox of valuable techniques and metrics, and even software packages that support them. I believe that ingenious tools such as the WBS, CPM, resource leveling, and earned value metrics offer enormous value. I also believe that most practitioners of project management are bright and conscientious.

But if project performance is erratic or poor, something must be lacking. What is it? The truth of that old management expression, that what is measured is what you get, suggests that project management needs to include measurement of new things. Starting to focus on, quantify, and maximize the business value of a project investment suddenly forces consideration of all the other factors of a project that affect that value:

- The details of the scope, which means those items that will generate the value
- The length of time before the project investment "matures" and starts to deliver its value
- How that value may be affected, positively or negatively, by external factors such as market window
- The resources that it takes to accomplish the scope and perhaps generate more value faster
- The cost of resources, which cuts into the investment's profit
- The risk that pruning the scope, compressing the time, or reducing the cost might interact with one of the other parameters to reduce our investment's expected profit by compromising quality, increasing cost through rework, or delaying project completion and the implementation of the project's value

Once the impact of these factors on one another and on the overall investment is quantified in monetary terms, items that were never previously monetized suddenly spring to prominence. They require techniques that are enhancements of the traditional project management toolbox:

1. If details of scope affect the project's expected value, then we need to quantify the value that optional work packages are adding. Each will certainly cost money and take time. We wouldn't want to perform a work package that adds less value than it costs. And thus the work breakdown structure, where we sum the cost of resources, should have as its value generation equivalent a value breakdown structure where we estimate each work package's value-added.
2. If the duration of the project will affect its expected value, then we need to manage that duration, which is driven by the critical path. But each item on the critical path contributes varying amounts to that duration, and so we need to know how much each item is delaying us. That leads to the crucial new CPM metric, critical path drag.
3. Critical path drag measures how much time each item on the critical path is adding to the project duration; but if a project is an investment, then we need to monetize that time. What we really need to know is how much that added time is reducing our project profit. That reduction of value comes from later delivery of the project's product or service, so we need to know, at the project level, what is the value/cost of time. That is probably the single most important item that could change the way we do projects: ceasing to allow the value/cost of time to be an unmeasured externality in terms of the value of the project investment. All it would take is two simple line items added to the business case of every project: how does the sponsor/customer value each unit of acceleration or delay in project completion?
4. Work activities can have critical path drag, but so can resource insufficiencies. We need to separate those schedule delays that are caused by the nature of the work from those that are caused by unavailable resources. The former is hard to change; but the latter might be prevented at much lower cost by changing an assignment or by increasing staffing levels.
5. Finally, although earned value metrics allow us to monitor cost trends with great accuracy (and schedule trends too, despite problematic distortions), once we recognize that a project is an investment, surely we will want to monitor its investment trends: the value it is expected to generate given the cost and schedule trends that we are already monitoring. And that is measured by the DIPP and the DPI, with the input from the earned value metrics that we already generate.

What we measure, we can and will improve. Once we start monitoring projects as investments, we will find ways to increase their value. Team members will start thinking outside the box, seeking ways to compile a DPI greater than 1.0. This would mean that the project will generate more value than was predicted. And shouldn't that be the goal of every member of the project team and every member of the organization?

Appendix

Exercises I: CPM, DIPP, drag, drag cost, and true cost

You are the project manager in charge of the project shown in Figure A.1. If the project is completed in 45 days, its expected monetary value will be $500,000. Every day later will cause a delay penalty of $50,000. Every day earlier will generate an acceleration premium of $20,000.

Lay out the network logic diagram for the project and answer the questions below. The answers are given, along with the network diagram, at the end of the Appendix.

1. What is the project duration with the current schedule as laid out in the diagram?

   _____

2. What is the starting DIPP with the current schedule as laid out in the diagram?

   _____

3. What would be the starting DIPP if the project duration were 45 days?

   _____

4. Which activities comprise the critical path?

   _____

5. What is the drag of each critical path activity?

   _____

6. What is the drag cost of each critical path activity?

   _____

7. What is the true cost of each activity?

   _____

Activity	Duration	Successors	Budget
A	8 Days	B,C	$10,000
B	12 Days	D	$20,000
C	10 Days	D,E	$30,000
D	4 Days	F,H	$20,000
E	6 Days	G	$30,000
F	10 Days	I	$40,000
G	7 Days	I	$20,000
H	13 Days	I	$20,000
I	10 Days	None	$10,000

Figure A.1 Chart of activities with successors, durations, and budgets for a CPM exercise.

8. What would be the impact on the project duration if Activity G takes 14 days?

_____

9. What would be the impact on the project duration if Activity B takes only 7 days?

_____

Exercises II: Business value and earned value

Goliath program

In the following scenario, you may assume that all future dollar amounts are already time and risk discounted.

You are a project manager working in the information systems department of a large company that specializes in consumer electronics. Your department has a new program called the Goliath Associated Systems Program (GASP). You have been made project manager of the first project in the program, to implement a new enterprise-wide platform called the Goliath Platform Project (GPP). As soon as your platform project is completed, it will permit global inventory control that is expected to save your company a net of $25,000 per week over the next 300 weeks after implementation. The Goliath Platform will also:

A. Enable the customer support department to implement a new software package that will allow for a 25% reduction in labor costs. This is expected to be completed 50 weeks after the Goliath Platform and to save a net of $100,000 per week over the next 250 weeks after implementation. This system is expected to cost $10M.

B. Enable the accounts receivables department to implement a new billing system to make collection of payments faster. This system is expected to be completed 50 weeks after the Goliath Platform and to generate a net of $20,000 per week in added revenues over the next 250 weeks after implementation, and will cost $1M to implement.

C. Enable the sales department to implement new software for identifying and facilitating sales leads. This system is expected to be completed 100 weeks after the Goliath Platform and increase sales by a net of $200,000 per week over the next 200 weeks after implementation, using a system that will cost $20M to implement.

Your project to implement the Goliath Platform Project has a budget of $3M and a planned project duration of 30 weeks. No acceleration premium or delay cost for your project has been specified, but you have been told that completion earlier would be helpful, as the inventory savings would start earlier. Later completion would reduce the amount of savings by a similar amount.

A. Project and Program Evaluation

Note: To answer these questions, you should have completed through Chapter 3 of the book.

A1. If each project finishes exactly on schedule, what will be the expected monetary value (EMV) of the entire program for the 300 weeks after Goliath has been implemented?

_____

A2. What will be the expected program profit?

_____

If each project finishes exactly on schedule, what will be the expected project profit on:

A3. Project A? _____

A4. Project B? _____

A5. Project C? _____

A6. Project Goliath? _____

If all projects are assumed to finish exactly on schedule, what will be the starting simple DIPP of:

A7. The entire program? _____

A8. Project A? _____

A9. Project B? _____

A10. Project C? _____

A11. Project Goliath? _____

B. Project Scheduling and the Value/Cost of Time

Note: To answer these questions, you should have completed through Chapter 6 of the book.

You have been told that the three other projects (A, B, C) will all start before the Goliath platform is completed, but that the Goliath platform will need to be completed at a certain point in the development of the other three systems. In addition, all the other projects would be accelerated to start generating value sooner if it were to become possible.

- An activity on the critical path of the Project A system implementation (with three weeks of drag) will need Goliath completed before it can start. This particular activity's only other internal predecessor has three weeks of total float.
- An activity that has one week of total float on the Project B system implementation will need Goliath completed before it can start. Its only other internal predecessor is on the critical path with one week of drag.
- An activity that has three weeks of total float on the Project C system implementation needs Goliath completed before it can start. It's only other internal predecessor is on the critical path and has three weeks of drag.

What will be the total acceleration premium (plus) or delay cost (minus) of Project Goliath if it finishes at the end of:

B1. Wk 26? _____

B2. Wk 27? _____

B3. Wk 28? _____

B4. Wk 29? _____

B5. Wk 30? _____

B6. Wk 31? _____

B7. Wk 32? _____

B8. Wk 33? _____

B9. Wk 34? _____

C. Earned Value, DIPP, and DPI

Note: To answer these questions, you should have completed through Chapter 9 of the book.

Your project to implement the Goliath platform has a budget of $3M and a planned project duration of 30 weeks. At the end of Week 10:

a. The PV is $1,000,000, the EV is $1,120,000, and the AC is $1,180,000.
b. No activities have slipped beyond their float, nor is there any indication that any will.
c. No risk factors have come into play.

C1. What is the current SPI?

_____

C2. What is the current CPI?

_____

C3. What is the current critical ratio CPI?

_____

C4. What is the current schedule EAC?

_____

C5. What is the current cost EAC (using the simple CPI)?

_____

C6. What is the current cost EAC (using the critical ratio CPI)?

_____

Note: For all the remaining questions, use the simple CPI.

C7. What is the current expected cost overrun/underrun?

_____

C8. What is the current cost ETC?

_____

C9.   What was the planned DIPP at Week 10?

_____

C10. What is the numerator of the current actual DIPP formula?

_____

C11. What is the current actual DIPP?

_____

C12. What is the current DIPP progress index (DPI)?

_____

C13. What is the increase/decrease in the expected project profit of Project Goliath?

_____

Answers to exercises I: CPM, DIPP, drag, drag cost, and true cost

The schedule should be as laid out in the network diagram in Figure A.2.

1. What is the project duration with the current schedule as laid out in the diagram?

   47 days.

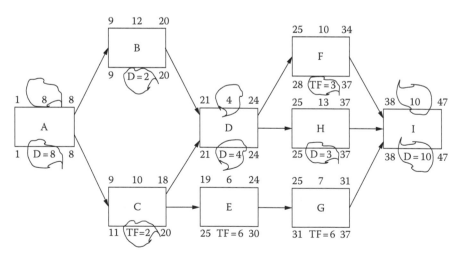

Figure A.2 CPM network diagram with drag computations.

2. What is the starting DIPP with the current schedule as laid out in the diagram?

($500,000 - $100,000) ÷ $200,000 = 2.0.

3. What would be the starting DIPP if the project duration were 45 days?

$500,000 ÷ $200,000 = 2.5.

4. Which activities comprise the critical path?

A – B – D – H – I.

5. What is the drag of each critical path activity?

A = 8, B = 2, D = 4, H = 3, I = 10.

6. What is the drag cost of each critical path activity?

A = (2 * $50,000) + (6 * $20,000) = $220,000; B = (2 * $50,000) = $100,000; D = (2 * $50,000) + (2 * $20,000) = $140,000; H = (2 * $50,000) + $20,000 = $120,000; I = (2 * $50,000) + (8 * $20,000) = $260,000.

7. What is the true cost of each activity?

True cost = budget + drag cost.
A = $10,000 + $220,000 = $230,000; B = $20,000 + $100,000 = $120,000; C = $30,000; D = $20,000 + $140,000 = $160,000; E = $30,000; F = $40,000; G = $20,000; H = $20,000 + $120,000 = $140,000; I = $10,000 + $260,000 = $270,000.

8. What would be the impact on the project duration if Activity G takes 14 days?

The project duration would be extended by 1 day to Day 48 due to Activity G taking 7 extra days and having only 6 days of total float.

9. What would be the impact on the project duration if Activity B takes only 7 days?

The project duration would be pulled in by 2 days to a duration of 45 days because Activity B has only 2 days of drag. The other 3 days by which Activity B's duration was compressed would now give Activity B total float of 3 days, and Activity C would now be on the critical path with drag of 3 days.

Answers to exercises II: Business value and earned value, the Goliath program (a)

A. Project and Program Evaluation

A1. If each project finishes exactly on schedule, what will be the expected monetary value (EMV) of the entire program?

($25K * 300) + ($100K * 250) + ($20K * 250) + ($200K * 200) = $77.5M.

A2. What will be the expected program profit?

$77.5M − ($3M + $10M + $1M + $20M) = $43.5M.

If each project finishes exactly on schedule, what will be the expected project profit on:

A3. Project A? ($100K * 250) − $10M = $25M − $10M = $15M.

A4. Project B? ($20K * 250) − $1M = 5M − $1M = $4M.

A5. Project C? ($200K * 200) − $20M = $40M − $20M = $20M.

A6. Project Goliath? ($25K * 300) + ($15M + $4M + $20M) − $3M = ($7.5M + $39M) − $3M = ($46.5M − $3M) = $43.5M (i.e., Project Goliath's revenues plus the profits it enables minus Project Goliath's budget).

If all projects are assumed to finish exactly on schedule, what will be the starting Simple DIPP of:

A7. The entire program? $77.5M/$34M = 2.28.

A8. Project A? $25M ÷ $10M = 2.5.

A9. Project B? $5M ÷ $1M = 5.0.

A10. Project C? $40M ÷ $20M = 2.0.

A11. Project Goliath? $46.5M ÷ $3M = 15.5.

Answers to exercises II: Business value and earned value, the Goliath program (b)

B. Project Scheduling and the Value/Cost of Time

What will be the total acceleration premium (plus) or delay cost (minus) of Project Goliath if it finishes at the end of:

B1. Wk 26? Accel premium of (4 * $25K) + (3 * $100K) = $100K + $300K = $400K.

B2. Wk 27? Accel premium of (3 * $25K) + (3 * $100K) = $75K + $300K = $375K.

B3. Wk 28? Accel premium of (2 * $25K) + (2 * $100K) = $50K + $200K = $250K.

B4. Wk 29? Accel premium of $25K + $100K = $125K.

B5. Wk 30? As planned. No acceleration premium or delay cost.

B6. Wk 31? Delay cost of −($25K + $100K) = - $125K.

B7. Wk 32? Delay cost of −([2 * $25K] + [2 * $100K] + $20K) = −$270K.

B8. Wk 33? Delay cost of −([3 * $25K] + [3 * $100K] + [2 * $20K]) = −$415K.

B9. Wk 34? Delay cost of −([4 * $25K] + [4 * $100K] + [3 * $20K] + $200K) = −$760K.

Answers to exercises II: Business value and earned value, the Goliath program (c)

C. Earned value, DIPP, and DPI

C1. What is the current SPI?

SPI = EV ÷ PV = $1,120,000 ÷ $1,000,000 = 1.12.

C2. What is the current CPI?

CPI = EV ÷ AC = $1,120,000 ÷ $1,180,000 = 0.95.

C3.   What is the current critical ratio CPI?

CR CPI = SPI * CPI = 1.12 * 0.95 = 1.06.

C4.   What is the current schedule EAC?

Schedule EAC = planned duration ÷ SPI = 30 weeks ÷ 1.12 = 26.78 weeks (27 weeks).

C5.   What is the current cost EAC (using the simple CPI)?

Cost EAC = budget ÷ CPI = $3,000,000 ÷ 0.95 = $3,157,895.

C6.   What is the current cost EAC (using the critical ratio CPI)?

Cost EAC = budget ÷ CR CPI = $3,000,000 ÷ 1.06 = $2,830,189 (i.e., by working ahead of schedule, we may be able to reduce marching army costs and finish below budget!).

Note: For all the remaining questions, use the simple CPI.

C7.   What is the current expected cost overrun/underrun?

Cost overrun = Cost EAC − budget = $3,157,895 − $3,000,000 = $157,895.

C8.   What is the current cost ETC?

Cost ETC = cost EAC − AC = $3,157,895 - $1,180,000 = $1,977,895.

C9.   What was the Planned DIPP at Week 10?

Planned DIPP = EMV ÷ (budget − PV) = $46.5M ÷ ($3M − $1M) = $46.5M ÷ $2M = 23.25.

C10. What is the numerator of the current Actual DIPP formula?

Actual DIPP numerator = ($EMV ± $accel premium/delay cost) = $46.5M + $375K (for 27-week estimated schedule at completion) = $46,875,000.

C11. What is the current Actual DIPP?

Actual DIPP = ($EMV ± $accel premium/delay cost) ÷ cost ETC = ($46.5M + $375K) ÷ $1,977,895 = $46,875,000 ÷ $1,977,895 = 23.70.

C12. What is the current DIPP Progress Index (DPI)?

DPI = actual DIPP ÷ planned DIPP = 23.70 ÷ 23.25 = 1.02.

C13. What is the increase/decrease in the expected project profit of Project Goliath?

Current EPP = EMV − cost EAC = \$46,875,000 − \$3,157,895 = \$43,770,105. Increase in EPP for Project Goliath = \$43,717,105 − \$43,500,000 = \$217,105. This is also the increase in EPP for the Goliath Associated Systems Program.

Glossary

A

ABCP See As Built Critical Path.

AC See Actual Cost.

Acceleration Premium See also Delay Cost. Increase in the value of the final product due to earlier completion or delivery date.

Activity-Driven Milestone A milestone representing the start or finish of an activity or work package. Used on US Department of Defense programs for proportioning earned value credit.

Actual Cost (AC) The realized cost incurred for the work performed on an activity during a specific time period. Referred to by the US Department of Defense as Actual Cost for Work Performed (ACWP).

ALAP Schedule See As Late As Possible Schedule.

Ancestor All activities that precede another activity on a logical path, including immediate predecessors, their predecessors, their predecessors' predecessors, and so on.

As Built Critical Path (ABCP) The actual critical path on the final schedule representing the completed work as performed. This is the critical path consisting of work, delays, and constraints that ultimately determines the final length of the project.

As Late As Possible Schedule (ALAP Schedule) Schedule of activities or milestones with all float squeezed out, at their late dates as computed by the backward pass of the critical path algorithm. This can provide an earned value baseline from which float distortions have been removed.

Availability Premium A cost to a project, program, or organization to guarantee that a specific resource will be available when needed. This can take the form of overstaffing or paying a fee to reserve the specified resource.

B

Backward Pass Algorithm See CPM Algorithm.

Basis of Estimates (BOE) Supporting documentation outlining the details used in establishing project estimates such as assumptions, constraints, level of detail, ranges, and confidence levels.

Bayesian Probability A probabilistic estimate that is updated as more relevant data become available.

BCWP See Earned Value.

Beta Distribution A Gaussian distribution of probability used in three-point estimating and Monte Carlo simulations to predict probabilities of project and activity duration, effort, and cost.

Billable Resources Resources in a contractor organization whose time is billable to a specific customer.

Black Holes Departments or other functional areas that are so under-staffed and overworked that not even information, such as estimates, can escape.

Black Swan A metaphor that describes a totally unanticipated event that has a major impact. Coined by Nassim Nicholas Taleb in his 2007 book of that title, black swans represent a major challenge for project risk planning.

BOE See Basis of Estimates.

Bottleneck A resource over-allocation during a period of time that makes the resource unavailable to a project and may delay project completion.

Budget Reserve Also called "cost reserve," this is additional funding available to a project to mitigate risk. Budget reserve and schedule reserve are two different types of management reserve, and are the property of the project manager or the customer/sponsor to be allotted as they deem appropriate.

Budgeted Cost for Work Performed (BCWP) The term used by the US Department of Defense for earned value. See Earned Value (EV).

Budgeted Cost for Work Scheduled (BCWS) The term used by the US Department of Defense for Planned Value (PV).

Burn Rate The rate at which a project or program is spending money per calendar unit or reporting period.

Burst Point The point in a schedule where one activity precedes two or more successors.

Business Value The value to an organization, program, or profit that any work effort is undertaken to generate.

C

CLUB See Cost of Leveling with Unresolved Bottlenecks.

Complex Dependencies Schedule relationships other than simple finish-to-start (FS). These can include start-to-start (SS), finish-to-finish (FF), and start-to-finish (SF) relationships, as well as lags.

Cost Estimate-at-Completion (Cost EAC) An estimate of what the project will have cost when it is completed. At the start of a project, its cost EAC is its budget. As the project progresses, the cost EAC is often computed by dividing the budget by the cost performance index (CPI).

Cost Estimate-to-Complete (Cost ETC) An estimate of what it will cost to complete a project from any given point. This is often computed by subtracting an ongoing project's actual cost from its cost estimate-at-completion (cost EAC).

Cost of Leveling with Unresolved Bottlenecks (CLUB) The cost to a project or program in reduced expected project profit (EPP) due to the delay caused by a specific resource bottleneck.

Cost Performance Index (CPI) This is an earned value metric used to perform cost trend analysis. It is a measure of the cost efficiency of budgeted resources expressed as the ratio of earned value to actual cost. The earned value formula for computing it is $CPI = EV \div AC$ (or in US Department of Defense terms, $CPI = BCWP \div ACWP$). In simplest terms, future over- or underspending is projected to follow the trend of what has happened thus far, so that the total budget is divided by the CPI to estimate the cost estimate-at-completion (cost EAC).

Cost Plus Contract A category of contract that involves payment to the contractor for all legitimate actual costs incurred for the completed work. Sometimes a fixed fee or an incentive based on the contractor meeting specific goals is added to the cost.

Cost Reserve See Budget Reserve.

Cost/Schedule Integration A system of techniques or software whereby the impact of cost modifications can be seen on the schedule and the impact of any schedule modifications can be seen on the cost.

Cost Variance (CV) The difference between what was budgeted for the work performed to any given point and what it actually cost. The earned value formula is: $CV = EV - AC$ (or in US Department of Defense terms, $CV = BCWP - ACWP$).

CPI See Cost Performance Index.

CPI-Labor A cost performance index generated by isolating labor hours or labor cost from the cost of other resources. It can be a very useful index for predicting future trends on projects that have large budgets for equipment and materials or where most of the labor is performed by exempt employees.

CPM Algorithm An algorithm in project management software packages that uses the sequence and durations of activities to compute the possible dates for each activity and to identify the longest (i.e., critical) path. The CPM algorithm actually consists of two separate algorithms: the forward pass algorithm that computes the earliest possible start and finish for each activity, and the backward pass algorithm that computes the latest possible start and finish for each activity without delaying the end of the project.

CPM Finish Date The earliest date that the project finishes based on the calculations of the CPM algorithm.

CPM Schedule The schedule for all activities generated by the CPM algorithm.

CPM Scheduling Generating a schedule based on the logical sequence and work of each activity.

Crash Duration An estimate of the least amount of time that an activity can take even if given unlimited resources.

Critical Path The sequence of activities that represents the longest path through a project, and which therefore determines the shortest possible duration. A program also often has a critical path, but is usually comprised of projects and based on value generation rather than physical logic.

Critical Path Drag The amount of time by which each item on the critical path (activity, constraint, or bottleneck) is delaying the end of the project or, alternatively, the amount of time by which the project schedule could be compressed by reducing the duration of any critical path item to zero.

Critical Ratio Cost Index (CRCI) An earned value formula for estimating a project's cost at completion, which recognizes that project delays will also add to the project cost. It is the product of both the CPI and the SPI: the CPI multiplied by the SPI. Cost EAC = budget ÷ (CPI * SPI).

CV See Cost Variance.

D

Damage Control Time Chart A chart originally developed by the author for use in optimizing planned schedules on emergency response projects. It shows negative impacts of the passage of time and the positive impacts available through project acceleration. It can be used for any type of project in any industry.

Date Ribbon The calendar periods listed horizontally at the top of a Gantt chart.

Deadline Originally, a real deadline: the line set up approximately 20 feet inside the fence of a US Civil War prison camp. Any prisoner venturing beyond that line would be shot by the guards. The term has come to be misapplied to target finish dates for projects and programs.

Default Distribution A common distribution shape used when running the algorithm of a Monte Carlo simulation program. The software permits the user to select a distribution for each activity from an agenda of many alternatives, but most users simply run it on one of the defaults: the triangular distribution, which uses the average of three inputs, or the beta distribution, which uses the PERT formula (or beta distribution).

Delay Cost See also Acceleration Premium. Decrease in the value of the final product due to later completion or delivery date.

Dependency See Predecessors.

Devaux's Index of Project Performance (DIPP) Originally an index for making decisions about when to abort projects, first published in the author's article, "When the DIPP Dips" in the September– October 1992 issue of *Project Management Journal*. In the Devaux book, *Total Project Control*, the DIPP was simplified for use as a tracking metric known as the Simple DIPP or the Tracking DIPP. That formula plans a baseline against which to track project progress in investment terms: Simple DIPP = ($EMV ± $acceleration premium/delay cost) ÷ $cost ETC.

DIPP See Devaux's Index of Project Performance.

DIPP Progress Index (DPI) A project investment metric that tracks business value against planned value: DPI = Actual DIPP ÷ Planned DIPP.

Doubled Resource Estimated Duration (DRED) An alternative developed by the author to the crash duration estimate. The DRED is a secondary duration estimate based on how long an activity would take if its resources were doubled. Used along with drag cost, it can be a useful tool for reducing an activity's true cost. See Resource Elasticity.

DPI See DIPP Progress Index.

Drag Cost The cost in reduced expected project profit at the time that a critical path item is adding to the project duration.

DRED See Doubled Resource Estimated Duration.

E

Early Dates The earliest dates that activities can start and finish based on CPM calculations.

Early Finish (EF) The earliest that an activity can finish based on CPM calculations.

Early Start (ES) The earliest that an activity can start based on CPM calculations.

Earned Schedule Tracking The relatively new earned value schedule tracking metric that translates the dollar units in which schedule variance is usually quoted into time units.

Earned Value (EV) The measure of work performed expressed in terms of the budget authorized for that work.

Earned Value Baseline See Planned Value (PV).

Earned Value Management (EVM) A methodology that combines scope, schedule, and resource measurements to assess project performance and progress.

Effort The number of labor units required to complete a schedule activity or work breakdown structure component, often expressed in hours, days, or weeks.

EMV See Expected Monetary Value.

Enabler Project A project part of whose value comes from enabling another project (often but not necessarily in the same program) to produce greater value.

Engineering Change Orders (ECOs) In US Department of Defense terminology, formal changes to the project or program baseline plan.

EVM See Earned Value Management.

Exempt Employee An employee who does not have to be paid for working overtime.

Expected Monetary Value (EMV) This is the whole purpose for which a sponsor/customer funds a project. It is the monetized business value that a project is expected to generate if it includes specific scope and finishes on a specific date. Factors beyond the responsibility of the project team can cause a project never to achieve its expected monetary value. However, the EMV should be a key metric in all project-based decisions. (Note that EMV is not EVM, which stands for earned value management and is a technique for analyzing project cost, not project value.)

Expected Project Profit (EPP) The expected monetary value of a project minus its cost.

Externality In economics, an externality is a cost or benefit that is not included in overall measurements and that may affect a party who had no say in its inclusion. For purposes of measurement, the impact of an externality is usually considered to be zero. In project management the value/cost of time is often left as an externality.

F

Fifth Edition of the *PMBOK® Guide* The *Guide to the Project Management Body of Knowledge* is published by the Project Management Institute (PMI) and is the authoritative source of current practice in the project management discipline. The fifth edition was published in 2013.

Fixed Price Contract An agreement setting the fee that will be paid for a defined scope of work regardless of the cost or effort to deliver it. A fixed price contract can also have an incentive fee that allows the contractor to earn an additional amount by meeting or surpassing certain criteria.

Float (slack) The amount of time that an activity's schedule can slip. See Total Float and Free Float.

Follow-On Value Value of a project that may come from other work or projects that it has enabled, such as revenues from a customization project for a system that was developed and sold to a client.

Free Float (Free Slack) The amount of time that an activity can slip without delaying the schedule of another activity.

Functional Department A department in an organization that is charged with certain functions and whose personnel have certain skills, such as mechanical engineering or documentation writing.

Functional Manager The head of a functional department who is usually charged with all personnel matters related to that department including assigning individuals to specific projects.

G

Gantt Chart Perhaps the most ubiquitous project management format, the Gantt chart was developed by Henry Lawrence Gantt at the Philadelphia Naval Shipyard in 1908. The date ribbon allows the user easily to see all work that is occurring simultaneously and resources that may be over-allocated (bottlenecks).

H

Hard Dependency This is a sequencing relationship between two items of work that is forced by the laws of physics: the wall cannot be papered until it's plastered and the software code cannot be debugged until it's written. It is very difficult for most project managers to find workarounds to the laws of physics.

I

Investment The outlay of money for the generation of greater value, usually, but not always, for revenue or profit.

K

Kahneman, Daniel 2002 Nobel laureate in economics. His 2011 book, *Thinking, Fast and Slow,* has significant implications for managing projects and explanations for why such management is often performed poorly.

Kindled Value This is the phenomenon where projects in a program often can kindle each other's value, that is, generate greater combined value than the value of each of the individual programs.

L

Lag In CPM scheduling, a planned delay between any two events.

Late Finish (LF) In CPM scheduling, the latest that any activity can finish without delaying the end of the project.

Late Start (LS) In CPM scheduling, the latest that any activity can start without delaying the end of the project.

Loss Aversion In Daniel Kahneman's book, *Thinking, Fast and Slow,* this is the tendency of human beings to be willing to pay more than it's worth to avoid a loss. It may explain why project managers and sponsor/customers are much more concerned about losses from finishing later than a target date than the value that they might generate by accelerating the schedule and finishing earlier, even if the delay cost and acceleration premium are identical.

M

Management Reserve A fund of extra time (schedule reserve) and money (budget reserve) set aside to mitigate risk and that the project manager and sponsor can use at their discretion.

Mandatory Activities Work activities that must be performed in order for the project to be completed. Mandatory activities have value in the VBS equal to that of the entire project as the project cannot be completed without them. See also Optional Activity.

Marching Army Costs Costs associated with supporting the project for as long as it takes to complete it. These can include project support costs, overhead costs, and opportunity costs. Shortening the duration of the project often has the positive by-product of reducing marching army costs.

Merge Points A point in the schedule where many predecessors merge into one successor. See also Burst Points.

Milestone A milestone is an event in the project, an instant in time entered into project management software such as an activity with a duration of zero. See Activity-Driven Milestones.

Milestone Weighting Milestone weighting is an earned value tracking technique that tries to remove some of the subjectivity of percent complete estimates by attaching earned value to activity-driven milestones instead of the activities themselves.

Monte Carlo Simulations A technique used to improve cost and schedule estimates by running probabilistic inputs with specific distribution functions thousands of times in order to see the likelihood of any one result.

N

Net Present Value (NPV) This is an estimate of how much value an investment will generate, taking into account risk and time factors. In this book, the term "expected monetary value" (EMV) is used as an alternative in order to express the fact clearly that the sponsor/customer of a project expects to achieve a certain value.

Net Value-Added The value of an activity or work package in a project taking into account the value it's expected to add minus its true cost.

O

Opportunity A project risk with a positive potential outcome.

Optional Activity An activity in a project that is not mandatory and could be left out without completely destroying the value of the project investment. An optional activity has value equal to the value of the project with it included minus the value of the project if it were omitted.

Order-to-Market In many industries, whether a product is first, second, third, or later to market is a major determinant of eventual market sales. This can have a major impact on a project's value/cost of time as an acceleration premium or delay cost.

Out-of-Sequence Work Work that is performed in contradiction of the logical schedule dependencies. This can foul up schedule analysis and can result in costly rework.

P

Pacing the Project Performing noncritical activities so that all paths finish at the same time as the critical path.

Pareto Chart A type of chart developed by Vilfredo Pareto often used in quality analysis to show in descending order the cost of different factors in an organization. This can be a very useful technique for prioritizing the cost of resource bottlenecks.

Parkinson's Law The principle expressed in 1955 that "Work expands so as to fill the time available for its completion."

Percent Complete A way of estimating progress on activities that suffers from being too subjective. To correct this, earned value tracking often relies on activity-driven milestones, which do not require percent complete estimates.

Perfectly Resource Elastic The quality of a small percentage of activities for which doubling the assigned resources will cut the duration to precisely 50% of what it was.

PERT This is an acronym that stands for program evaluation and review technique. It was developed by consultants at the consulting firm Booz Allen Hamilton in 1958 for use on the US Navy's Polaris missile program. It uses a weighted formula for estimating activity effort and durations using three estimates for each activity: optimistic, most likely, and pessimistic.

PERT Chart This terminology is used today simply to mean what is more properly called a project network logic diagram. When someone asks to see a PERT chart, they are usually not expecting use of the PERT formula.

Planned Cost A cumulative function within a project based on the budget accruals for the activities. In earned value tracking, this function is what is termed the planned value (PV) function, or the earned value baseline. In US Department of Defense terminology, this would be the budgeted cost for work scheduled (BCWS).

Planned Cost Accruals See Planned Value (PV). How project costs are expected to accumulate over time.

Planned Cost Estimate-to-Complete (ETC) What it is expected to cost to complete the project from any date forward during the project. At the start of the project, the planned cost ETC is identical to the budget. But as the scheduled work is performed and paid for, the planned cost ETC should decline by the amount that was budgeted for the work already performed (i.e., as the complement of the planned value function).

Planned Value (PV) The baseline for earned value tracking, it is important to remember that it is a cost function, not a value function. It is the cumulative budgets for all activities as scheduled. In US Department of Defense terminology, it is called the budgeted cost for work scheduled (BCWS).

PMI See Project Management Institute.

Predecessors Also called logical dependencies, these are the activities that come immediately before other activities. They are the most proximate of an activity's ancestors or dependencies.

Prime Operating Metric The guiding quantitative measure of an endeavor. For projects, as for all other investments, the prime operating metric should be profit, or value-above-cost.

Principal–Agent Problem This is an issue in economics involving the difficulty in motivating the party in an endeavor or contract that is closest to the effort to act in the best interests of the other rather than in the best interests of themselves. This can be a major problem when projects are performed on a contractual basis and the benefits to the customer and the contractor have not been properly aligned.

Product Scope The features and functions that characterize the product, service, or result of a project.

Program A group of related projects, subprograms, and program activities managed in a coordinated way to obtain benefits not available from managing them individually.

Project In the *PMBOK Guide* terminology, a temporary endeavor to create a unique product, service, or result. This definition unfortunately does not include the vital information that every project is an investment. Perhaps a better definition would be: "An investment in work to create a unique product, service, or result."

Project Business Case This should be part of the documentation during the initiation phase of a project. It should lay out the reason for the project, what its value is expected to be, and how that value may be affected by factors such as project duration.

Project Control The processes by which a sponsor/customer or other stakeholder can ensure that a project is being performed in a way that will achieve its important goals. Because a project is an investment, the most important of these goals that should be controlled is value above-cost, our project profit.

Project Management Institute (PMI) An organization of professionals in the project management discipline. It is based in the United States but has chapters throughout the world. It is responsible for several certification programs in the discipline and for publishing periodic updates to the *Guide to the Project Management Body of Knowledge*.

Project Management Journal The refereed magazine of the Project Management Institute (PMI). It was in this journal that the author's article, "When the DIPP Dips," introduced the concept of the DIPP in September–October, 1992.

Project Profit The difference between a project's cost and its value. Because every project is an investment, projects are only funded if they're expected to produce a profit in this sense.

Project Scope The work performed to deliver a product, service, or result with the specified features and functions. Sometimes referred to as "work scope."

Project Sponsor The individual who provides the resources to do a project. The sponsor can also be considered the project investor and is also usually the person who expects to benefit from the value that the project generates.

Purchased Items Distortion In earned value tracking, the tendency of costs for equipment and materials to far outweigh labor costs despite the fact that labor-intensive activities tend to be riskier and more volatile.

Q

Quality The degree to which a set of inherent characteristics fulfills requirements.

R

RAD See Resource Availability Drag.

Remaining Duration The amount of time necessary to finish an activity that has already started.

Resource Availability Drag (RAD) The delay in the project schedule due to a resource bottleneck or other unavailability of a specific resource when it is needed.

Resource Availability Drag Cost The cost in reduced expected project profit due to a schedule delay caused by the lack of availability of a specific resource. See also Cost of Leveling with Unresolved Bottlenecks (CLUB).

Resource Elasticity The tendency of an activity's duration to expand or contract in response to increases and decreases in resource levels. See also Doubled Resource Estimated Duration (DRED).

Resource Leveling The process, supported by many project management software packages, to even out resource usage across a project schedule. The two key resource leveling processes are time-limited resource leveling and resource-limited resource leveling.

Resource Library The database of all the resources available to projects in an organization and their usage and availability shown on a calendar.

Resource Limited Resource Leveling The process to even out resource usage and limit it to the availability of resources as reflected in the resource library. This process often requires delaying the scheduled project completion in order to live within the limits of resource availability.

Resource Limited Resource Schedule The schedule produced by resource-limited resource leveling.

Resource Over-Allocations See Bottleneck.

Risk Retirement The process of recognizing that an identified risk factor is no longer relevant. A project plan should include dates by which if a risk factor has not become manifest, the risk and any mitigating tactics associated with it will be retired and no longer affect the project's expected monetary value.

S

S-curves The three tracking functions of earned value (PV, EV, and AC) as plotted on a schedule.

Schedule Performance Index (SPI) An earned value metric used to perform schedule trend analysis: SPI = EV ÷ PV. The schedule performance index suffers from the fact that, as usually implemented, it does not recognize the critical path or its crucial importance to the project schedule.

Schedule Reserve Management reserve in the form of extra time in the schedule.

Schedule Variance (SV) Schedule variance is simply the difference between what was budgeted for the work that has actually been completed (EV) and what was budgeted for the work that was scheduled to have been completed thus far (PV). The formula is: SV = EV − PV. Like the schedule performance index it suffers from the fact that, as usually implemented, it does not recognize the critical path or its crucial importance to the project schedule.

Simple DIPP A simplified version of the DIPP as originally published in the *Project Management Journal* that can be created as a baseline and tracked during project execution. The formula for the simple DIPP is: ($EMV ± $acceleration premium/delay cost) ÷ $cost ETC. Also known as the tracking DIPP.

Slack See Float.

Soft Dependency Also known as a discretionary dependency, this is a predecessor–successor relationship input to the schedule on the basis of a project manager's decision rather than on the basis of the logic of the physical work. If the soft dependency is on the critical path, then the drag and drag cost should be computed to ensure that such a dependency is not costing more than it's worth.

SPI See Schedule Performance Index.

Sponsor/Customer The individual, individuals, or organizations that initiate a project and invest in it and hope to reap the value that it generates.

Successor Activity An activity that comes immediately after another activity. Successor activities are the most proximate descendant activities to any given predecessor.

T

Thinking, Fast and Slow A 2011 book by 2002 Nobel laureate Daniel Kahneman that may explain many of the reasons that project management decisions are often made that are unjustified from an economics point of view.

Time & Material Contract (T&M) A contract between customer and contractor that typically pays the contractor a fixed rate for all material provided in the time consumed. This type of contract can create moral hazard for the contractor if there is no incentive or penalty clause attached to date of delivery.

Time-Limited Resource Leveling The process to even out resource usage and limit it to the availability of resources as reflected in the resource library. In time-limited resource leveling, the project completion date is fixed. If the date that is input is the CPM finish date, then no activity can be delayed beyond its total float and the schedule will continue to reflect resource bottlenecks that will have to be resolved.

To Complete Performance Index (TCPI) This is an earned value index during project execution that compares how much we have been achieving for the portion of the earned value we have achieved thus far and how much money is left in the budget to complete the rest of the work. It shows the level of the CPI that must be maintained for the rest of the project in order for it to finish on budget. The formula is: TCPI = (Budget-at-Completion – EV) ÷ (Budget-at-Completion – AC).

Too Many Cooks Syndrome A situation on a project where adding resources causes an activity to take longer. The DRED on such activities is higher than the original estimate.

Total Float (Total Slack) The amount of time that an activity's schedule can slip without delaying the end of the project.

Tracking DIPP See Simple DIPP.

Triangular Distribution One of the standard distribution shapes that a user can select for the duration of all the project activities to be scheduled in a Monte Carlo simulation software package. The triangular distribution, which is based on the input of lowest, most likely, and highest estimates, is the most common distribution shape used. The second most common distribution shape used is what is generally called the beta distribution, which reflects the PERT formula.

True Cost (TC) The true cost of doing work in a project is due both to the cost of resources for that activity and the impact of that activity's duration on the project schedule. The formula is: TC = resource cost + drag cost. The net value-added of an activity is its value-added minus its true cost. Most projects do not compute the drag cost of activities, therefore they often wind up including work that has a negative net value-added.

U

Utilization Rate This is the percentage of time that a resource or resources is billable to a specific customer or contract. Functional managers are often measured on their ability to maintain a high utilization rate. Unfortunately, this can present moral hazard as the actions that will keep a department's utilization rate high, such as multitasking and minimizing headcount, are often detrimental to the project investment by causing resource delays.

V

Value Breakdown Structure (VBS) A concept introduced in the author's 1999 book, *Total Project Control*, that separates a project's work packages and activities into mandatory and optional categories and estimates the value-added for each optional activity.

Value Drivers The factors that can generate value from any given project. Expected revenues and savings are the two most easily identified value drivers, but others such as follow-on business, customer satisfaction, patents, market dominance, and giving engineers experience with valuable new technologies can also drive significant value on certain projects.

Value-Added The difference between the value of a project with or without a specific work package or activity.

VBS See Value Breakdown Structure.

Index

Printed and bound by CPI Group (UK) Ltd, Croydon, CR0 4YY

28/10/2024

01780125-0001